PARALLEL MINDS

LAURA TRIPALDI

PARALLEL MINDS

Discovering the Intelligence
of Materials

URBANOMIC

Published in 2022 by

URBANOMIC MEDIA LTD.
THE OLD LEMONADE FACTORY
WINDSOR QUARRY
FALMOUTH TR11 3EX
UNITED KINGDOM

Originally published as
Menti Parallele © effequ 2021

English translation © Urbanomic Media Ltd. 2022

BRITISH LIBRARY CATALOGUING-IN-PUBLICATION DATA
A full catalogue record of this book is available
from the British Library

ISBN 978-1-913029-93-7

Image Research: Elaine Tam

Printed and bound in the UK
by Clays Ltd, Elcograf S.p.A.

Distributed by the MIT Press,
Cambridge, Massachusetts and London, England

CONTENTS

Foreword   *Matteo Giuli*   vii

Introduction   1

1. ARACHNE'S WEB   7
The Missing Majority 9; A Spider's Work 15; Structure
and Function 23; Weaving the Future 29

2. A MANY-HEADED THING   33
Hydra 35; Intelligent Jelly 41; Under the Skin 47; Being
in the World 53; Through the Looking-Glass 59

3. THE PATTERN WHICH CONNECTS   63
Golem 65; Thinking Complexity 71; Bricks and Atoms
77; Synthesising Complexity 87; Back to the Mind 99

4. LIVING MONSTERS   103
Artificial Lives 105; Inorganic Organisms 113; Other
Forms of Lyfe 121; Life and Information 127; The Prom-
ises of Monsters 131

5. WHAT THE FUTURE IS MADE OF   141
Minds in the Web 143; Arachne 2.0 149; Weavers of the
Future 155; Ariadne's Thread 165

Acknowledgements   169
Bibliography   171
Index   177

Foreword
Matteo Giuli

During the first year of my physics course, a certain professor who had just given an unnecessarily complicated explanation of one of the fundamental theorems covered in the semester was astonished at the many hands raised at the end of the lecture, and asked in mocking tones whether, given our obvious hopeless stupidity, we hadn't wandered in by accident from the 'Department of Romance Philology'. Almost everybody in the class laughed, but it was true that, by this point, it had already been drummed into us for some time that knowledge of the *world* is gained exclusively via knowledge of the *physical world*. Other disciplines may be worthwhile but, however interesting they may be, they play an inevitably secondary role: after all, what is engineering except physics without ideas, what is mathematics but physics without guts—and chemistry is just an obsolete physics, left in the dust by the twentieth century, while philosophy is a physics without rigour, for those who are too lazy to do calculations. And as for the human sciences, if you really must mention them, at least have the decency to use the obligatory air quotes when mouthing the word 'sciences'....

Such is the daily drip, drip, drip of reductionism—that widespread prejudice which, to quote the book you are about to read, posits 'a sort of hierarchy of scientific disciplines, in which physics is more fundamental than chemistry, which is in turn more fundamental than biology, and so on down to the human and social sciences'.

The passage quoted above is not central to *Parallel Minds*, yet it seems to me decisive in explaining what kind of essay you have before you here: a vital antidote to the asphyxia of reductionism and blind specialisation, a book that thrives on lateral ideas and alternative imagery, that is fuelled by hybrids and relationships: in between disciplines, living beings, objects, and technologies.

Parallel Minds invokes the ancestral spirit of the material sciences ('perhaps we cannot rule out the possibility that the lairs of the ancient alchemists were, among other things, primitive nanotech laboratories...'), weaves together mythological allegories (Ariadne, Icarus, Arachne) with stories of revolutionary scientific research, imagines the future of technology as a cultural challenge—a question of material culture—and develops its own hypotheses through analyses of the work of philosophers who have confronted the same challenges (if I have managed to keep track of them all: Sadie Plant, Peter Godfrey-Smith, John Searle, Maurice Merleau-Ponty, Andy Clark, Karen Barad, Edgar Morin, Isabelle Stengers, Jacques Monod, Donna Haraway, David Chalmers, Luce Irigaray, Jane Bennett).

In a 1944 letter, Albert Einstein used the perfect metaphor to describe the hyper-specialised types we tend to find working in science ('the majority', he wrote at the time), those who incarcerate themselves within the confines of their respective disciplines and refuse to consider their profession from any other angle: they are, he says, like 'somebody who has seen thousands of trees but has never seen a forest'.

As Einstein knew, only a broad culture can afford us a complete view of things: only 'a knowledge of the historic and philosophical background gives that kind of independence from prejudices of his generation from which most scientists are suffering'. And that this independence, where it is achieved, may be considered 'the mark of distinction between a mere artisan or specialist and a real seeker after truth'.

A few years ago, a number of scientists (including Carlo Rovelli and Alberto Mantovani) co-signed an article entitled 'Why Science needs Philosophy', which opened with an excerpt from this letter of Einstein's and closed on a sentence by the visionary and unorthodox biologist Carl Woese, a quote that is also a perfect fit with Laura Tripaldi's book:

> [A] society that permits biology to become an engineering discipline, that allows science to slip into the role of changing the living world without trying to understand it, is a danger to itself.[1]

---

1.   L. Laplane et al., 'Opinion: Why Science Needs Philosophy', *PNAS* 116:10 (5 March 2019), 3948–52.

The universe into which this book will take you is a perturbing, strange, unfamiliar territory, which is precisely why it is so fascinating. An astonishing place where a slime mould, without brains or sense organs, still manages to build up a representation of its environment and colonise it; where tiny particles of lifeless plastic material seem to behave with the conscious coordination of an anthill. And here you will also rediscover yourself, entangled in the physical space of reciprocal interaction between organisms and materials, within that wide network of exchanges between things in which we human beings are only one node among many—at which point you will realise with surprise that you no longer have any secure, comfortable definitions for fundamental and apparently self-evident concepts such as life and intelligence (and consciousness, and thought).

Parallel Minds is a rigorous scientific guide to complexity, but one with the intoxicating force of a manual of white magic. Its lucid gaze is capable not only of seeing Einstein's forest but of describing it to us, spelling out the hyper-connected, rhizomatic thoughts of the trees that make it up.

Introduction

> *Those were the days when the earth itself fornicated with the
> sky, when everything germinated and everything was fruitful.
> Not only every marriage but every union, every contact, every
> encounter, even fleeting, even between different species, even
> between beasts and stones, even between plants and stones, was
> fertile, and produced offspring not in a few months but in a
> few days. The sea of warm mud, which concealed the earth's
> cold, prudish face, was one boundless nuptial bed, all its re-
> cesses boiling over with desire and teeming with jubilant germs.*

Primo Levi, 'Quaestio de centauris'[1]

In his youth, before he became a celebrated writer, Primo Levi used to
work as a chemist in the manufactory of paints and varnishes. As he
explained on several occasions, he was particularly fond of the figure
of the centaur, an ambiguous monster in which two apparently incompatible bodies are merged. For Levi, these two parts stood above all for
the incompatible worlds of science and literature, the art of chemistry
and the art of storytelling. But the centaur, in its duality, also has much
to tell us about the complex and fecund encounter between the minds
of human beings and the untameable matter that surrounds them.

Today the word 'interface' has entered into common usage, but in
a slightly different sense from the one it has in chemistry and materials
science. We often use it to describe our interactions with new digital
technologies: we talk about application interfaces and the interfaces
of software or websites, to refer to the 'face' that technology presents
to us when it has to communicate with us. In this sense, the interface
is a window that opens up before our eyes, allowing us comfortably
to access parallel worlds otherwise inaccessible to us. It is the female
voice of the virtual assistant who guides us in our daily life, the social

---

1.   Tr. J. McPhee, *The New Yorker*, 8/15 June 2015.

network that asks us amicably what we're thinking about. It is the way
we make our technologies more and more human and immediate, often
by hiding the most contentious and complex aspects of their operations, neatly sweeping them under the carpet. This familiarity with the
interface, which we try to make ever thinner to the point of invisibility,
tends to make us forget that any dialogue with technology takes place
on a hybrid territory, where our instruments influence our behaviour as
much as we influence theirs.

Through working with materials, and having come face to face with
numerous occasions upon which communication between two surfaces has turned out to be more complex than expected, I have realised
that *interface* is a deeper and broader concept than it may seem at first
glance. If I had to pick out just one thing I have learned, among all the
surprising things I have had the opportunity to discover while studying chemistry, it would definitely be this: that the interface is not an imaginary line that divides bodies from each other, but rather a *material
region*, a marginal area with its own mass and thickness, characterised
by properties that make it radically different from the bodies whose
encounter produces it.

Anyone who is confronted with a new material soon realises that
what determines its behaviour often has nothing to do with its internal composition or structure, which in chemistry we call *bulk*, but with
what happens on its surface. The important thing is what happens in
the region where the encounter *between that material and something else*
takes place—an encounter which may be simple but is more often a
complicated affair. The interface, in chemical terms, is defined precisely as the region where two substances with different physicochemical
properties meet.

I remember very distinctly a moment some years ago when, while
working on my thesis, I was trying to deposit a very fine layer of tita-
nium dioxide nanoparticles onto a polymeric support—in other words,
a plastic disc of about ten centimetres in diameter floating on the surface of the water, which was meant to capture and remove pollutants
from the water, and, with the help of sunlight, break them down. The
nanoparticles and the polymer really didn't want to get along: when

the disc was submerged in water, the layer of particles detached themselves from the support, dispersing into the water and making it impossible for me to keep them together. A more familiar example of the effect of the interface on the properties of materials can be observed in the behaviour of water. A drop of water deposited on a sheet of glass, in contact with the air around it, naturally produces a semi-spherical surface. This phenomenon owes to the *surface tension* of water, a quantity that indicates the tendency of a substance's molecules to cohere with one another, reducing the surface area they present to the outside world. Interestingly, the behaviour of the drop is not simply an intrinsic property of water, but changes according to the characteristics of the different substances with which it interacts. So, for example, if the surface of the glass is chemically modified, depending on the type of molecules bonded to its surface, the drop of water will tend to flatten completely or, in other cases, to reduce its area of contact with the glass and form a perfect sphere. If, on the other hand, particular organic substances called *surfactants* are dissolved in the water and allowed to spread across the outermost surface of the drop, reducing its surface tension, it will show a greater affinity with the air and will prefer to flatten, exposing a greater surface.

These simple behaviours show that the interface is truly a space of encounter in which two different bodies come together to form a completely new state of matter. Even though the molecules of the water drop never change their chemical nature, within the interface they behave very differently than they usually do, arranging themselves into a particular structure depending on which substance they come into contact with. In this sense, the interface is the product of a two-way relationship in which two bodies in reciprocal interaction merge to form a hybrid material that is different from its component parts. Even more significant is the fact that the interface is not an exception: it is not a behaviour of matter observed only under specific, rare conditions. On the contrary, in our experience of the materials around us, we only ever deal with the interface they construct with us. We only ever touch the surface of things, but it is a three-dimensional and dynamic surface, capable of penetrating both the object before us and the inside of our own bodies.

This idea of the interface as a material region in which two substances can mix together to produce a completely new hybrid body, can serve as the starting point for rethinking more generally our relationship to the matter around us. If all the bodies we enter into relations with are modified and modify us in turn, then we can no longer delude ourselves that matter is simply a passive object onto which we project our knowledge. But neither can we take refuge in the convenient idea that we can never have any knowledge of that which is not human—that the matter around us is, ultimately, completely alien and unknowable, and that it really has nothing to do with us. Inhabiting the interface affords us the opportunity to redefine our knowledge of matter as a creative and collaborative process in which every material actively participates. Every time we enter into a relationship with a new material, we construct a physical space of mutual interaction which modifies the world around us and opens us to the possibility of modifying ourselves in turn.

In this context, what we have become accustomed to considering as the simple objects of science seem to come alive, instead becoming real subjects, taking an active part in the scientific process that defines and studies them. When seen from this new perspective, even bodies that we have always considered inert and passive reveal a latent capacity to weave a network of relationships with us and with the world around them. This is a conceptual path which, in recent years, has been followed in a number of different areas of scientific knowledge. While for a long time the belief that human beings had a sort of monopoly on intelligence has predominated, this new scientific outlook has enabled us to discover that not only animals closer to us such as mammals, but also invertebrate organisms, plants, and fungi are in fact subjects at the centre of a very rich perceptual and relational universe—an insight which radically challenges our idea of what a mind is. Many of these subjects have horizontal and delocalised minds: they are capable of thinking not with a specific organ, but with their whole body, and even outside the confines of their own organism. But discovering the intelligence of materials is not just a conceptual exercise, aimed at extending the notion of intelligence to the domain of non-living (in the strict sense) matter. On the contrary, investigating these material minds means above all trying to find the common root of all intelligences in the intrinsic vitality of the matter from which they are made.

Its creative approach to the study of matter is perhaps the thing which, more than anything else, convinced me to begin and to continue my study of chemistry, and later materials science. These two disciplines do not have a straightforward relationship, even if they start from a common ground. Where chemistry is often focused on the study of the properties and reactions of individual molecules, materials science looks at how the molecular structure of a body influences its macroscopic properties, i.e. those we deal with in everyday life: for example, mechanical resistance or the ability to conduct heat. In other words, materials science tries to understand how a large number of molecules, atoms, and many different microscopic components interact with one another to form complex structures with emergent properties which their simple elements alone do not possess. In this book I will talk about materials in reference to both chemistry and materials science depending on the context, i.e. depending on whether the focus is mainly on their molecular structure or on their macroscopic behaviour.

One of the most interesting and perhaps unique features shared by chemistry and materials science is the fact that these sciences can never really be separated from the technologies they produce. We usually think of science and technology as two strictly distinct moments: in many disciplines—in theoretical physics, for example—the discovery of the principles that determine the behaviour of bodies precedes, both conceptually and chronologically, their technological application. But the discovery of a new material necessarily coincides with the creation of a new technology: it is the production of a body that did not exist before. For, unlike other sciences, chemistry and materials science turn to the *synthesis* of new substances in order to achieve a deeper understanding of the potential of the material around us. What has always fascinated me about chemical synthesis is that here the process of synthesis is never unidirectional, but always *discursive*: the material that is studied becomes an active participant in the cognitive processes of the science that studies it and, within the fertile space of this interface, helps produce something new.

It is perhaps for this same reason that a broader reflection on the epistemological implications of these applied sciences is so often lacking. These are disciplines which undermine the anthropocentric paradigm of the natural sciences which always places scientists at a safe distance from the object they study, reinforcing the rigid distinction between human and nature. On the contrary, a synthetic approach to the study of matter continually renegotiates the categories we make use of to know the world: synthesis always confronts us with a substance which is hybrid, neither natural nor artificial, a material which is the tangible result of the interaction of matter with ourselves. The complexity of this relational process is also reflected in the intricate history of these sciences, which, to a great extent, is not a linear narrative but a labyrinth of concepts and procedures that have developed through everyday experimental encounters with matter in the laboratory. However, precisely because these disciplines contribute directly to shaping the technologies we use, and which will necessarily be those with which we face the challenges of the future, it is becoming increasingly urgent that we develop a shared language that allows us to include them in the cultural debate on science, while doing justice to their complexity. Although they have often been set aside as a matter of mere technical knowledge, perhaps these *unnatural sciences* can allow us to imagine a more horizontal, more open science, a science that would be less hierarchical and better integrated with the complex reality that surrounds it.

This book is about the weird encounters that take place within the interface. The materials I describe in these pages are the product of the complex and reciprocal interweaving of the intelligence of matter with our own: they are not the *objects* of a science, but are, to all intents and purposes, its sole protagonists. Perhaps, like the centaur, they too are monsters: hybrid creatures that inhabit the edgelands between what is human and what is not. Our mythologies, both ancient and contemporary, offer numerous examples of creatures whose ambiguous and unnatural bodies have served as so many occasions for us to renegotiate our own position in the world. In these creatures, daughters of the interface, we may discover new allies for the future that awaits us.

1

ARACHNE'S WEB

Then, going off, she sprinkled her with juice,
Which leaves of baneful aconite produce.
Touch'd with the pois'nous drug, her flowing hair
Fell to the ground, and left her temples bare;
Her usual features vanish'd from their place,
Her body lessen'd all, but most her face.
Her slender fingers, hanging on each side
With many joynts, the use of legs supply'd:
A spider's bag the rest, from which she gives
A thread, and still by constant weaving lives.

Ovid, *Metamorphoses*

Scanning electron microscope image from the Museum National d'Histoire Naturelle in Paris, showing multiple fibres in a Palaeolithic cord fragment. From Bruce Hardy et al., 'Direct Evidence of Neanderthal Fibre Technology and its Cognitive and Behavioral Implications'.

The Missing Majority

When we think about the history of technology, weaving is probably not among the first things that come to mind. And yet weaving has had an incalculable impact upon human civilisation: from the production of clothing to the birth of modern programming, this technology has accompanied human history as a silent presence, intertwined with the life of every one of us. The oldest traces of woven fibres among *homo sapiens* are known to date back more than twenty thousand years, to a time well before the birth of agriculture, but it is probable that our species was not the only one to have developed the technique of weaving: a recent study has revealed the discovery of a woven fragment attributed to Neanderthal man, which would take weaving back as far as ninety thousand years.[1]

Elizabeth Wayland Barber, an archaeologist and expert in the history of weaving, argues that widespread ignorance of this fundamental aspect of the history of technology owes largely to the perishability of fabric fibres, physical traces of which are easily lost with the passage of millennia.[2] This then is also the reason why, when we imagine technologies of the past, we think of hard materials such as stone and metal, while fabric, by its nature soft and organic, ends up being almost completely forgotten.

It is most likely that the 'Stone Age' was not really the Stone Age, and that most of the technologies used by prehistoric humans were characterised by the use of organic materials that have been almost completely lost. These perishable technologies that were certainly

1.   B.L. Hardy et al., 'Direct Evidence of Neanderthal Fibre Technology and Its Cognitive and Behavioral Implications', *Scientific Reports* 10 (2020), 4889.
2.   E. Wayland Barber, *Women's Work, The First 20,000 Years: Women, Cloth, and Society in Early Times* (New York: Norton, 1995).

a part of human life since the Palaeolithic, but of which only a
few traces remain, have been dubbed by the archaeologist Linda
Hurcombe *the missing majority*.[3] Hurcombe argues that this missing
technological knowledge is not just a problem of the incomplete-
ness of the archaeological record, but has also had an incalculable
influence upon our vision of past societies. The selective forgetting
of the past, so evident in the case of perishable materials used in
antiquity, has much to teach us about the way we think about the
technologies of the present and imagine the technologies of the
future. Because the materials we use are not passive objects but, on
the contrary, are determined by our socio-cultural life and in turn
determine our relationship with the world, forming what is called
material culture: a culture that is shaped by the invention, production, and use of the materials around us. In other words, a culture
cannot be separated from the materials that characterise it; when a
substantial portion of them is forgotten, our knowledge of that culture
suffers enormously.

We must always keep in mind that the materials we use on a daily
basis say a lot about our culture, and that our cultural perspective is a
determining factor in the choice of materials with which we build our
world. When we look at a fabric, we usually don't see it as a technological object, because its flexibility and softness do not fit our mechanistic
image of technology based on rigid, hard materials that can survive for
tens of thousands of years; yet in terms of complexity and adaptability,
fabric is a far more advanced material than a piece of metal. The same
prejudice that has influenced our perception of prehistoric technologies as nothing but a collection of sharp stones—and has made us ne-
glect fine weaving, food preservation, and pigment preparation—also
comes into play when we imagine a future based on steel and silicon.
It is possible, indeed it is certain, that we will have to learn to make our
technologies softer and more flexible if we are to have any chance of
overcoming the challenges that lie ahead. At a time when we are obliged
to reflect upon our impact on the planet, an impact so great that it

3.   L.M. Hurcombe, *Perishable Material Culture in Prehistory: Investigating the Missing
Majority* (London: Routledge, 2014).

has become geological, our best technologies will be those that leave no trace.

Perhaps another reason why we so rarely consider weaving is that it is an essentially feminine technology. A very common prejudice would have it that the techniques women have developed and kept alive since the dawn of our civilisations are not real technologies; instead they are treated almost as inexhaustible and mysterious natural resources: they are seen as the result of innate tendencies, their existence is taken for granted and, too often, their complexity is underestimated. The Jacquard loom, designed in 1801 by Jean-Marie Jacquard, is widely regarded as the first programmable machine ever designed, and used a system based on perforated cards surprisingly similar to those used more than a century later in the first computers. It was the Jacquard loom that inspired the mathematics of Ada Lovelace, who, together with Charles Babbage, developed the project for a computational machine called the *Analytical Engine,* in principle capable of performing any algebraic work. As is often the case with new technologies, the Jacquard loom also met with a great deal of resistance from public opinion. In a warning against its widespread adoption, the poet Lord Byron described weaving with the Jacquard machine as *spider-work.* Most likely Byron had never looked at a spider's web closely enough: if he had, he would have appreciated its extraordinary complexity, and perhaps would have learned to appreciate the loom's ingenuity too.

This digression on the most ancient art of weaving in a book dedicated to contemporary materials science may seem perplexing. It could be argued that materials science, unlike weaving, is a very recent technology, born and raised in the clinical white laboratories of the twentieth century and certainly not in the tepid half-light of a palaeolithic cave. What I find most striking about fabric, however, is not so different from what fascinates me when I look closely at a new material. At the basis of weaving and of modern materials science alike there lies the idea, as simple as it is revolutionary, that the repetition of many small identical elements—the 'threads' of the loom—can produce a new object with remarkable properties not possessed by the individual starting elements alone.

The transition from a simple tangled ball of yarn to a complex fabric is determined by *structure*: as every expert weaver knows well, weaving is not just any old way to hold threads together, but a set of specific methods which produce particular properties unique to each fabric, making it more elastic, more resistant, warmer, or more breathable. Many advanced materials today are designed and constructed in exactly the same way, trying to capture the exact convergence of material and structure—'thread' and 'weave'—capable of endowing the material with the desired properties. The fact that the loom was the model for modern computational machines highlights a fundamental feature of complex matter, namely that its structure can store a certain amount of information which is not imparted from the outside, like words inscribed onto a piece of paper, but is 'written' in the relationships between the microscopic elements that make it up. In other words, as in the case of fabric, the strength of our most innovative materials lies in their *cooperative* and *relational* nature, which is expressed in a diffuse and decentralised structure endowing the material with properties that its individual parts do not possess. The result is an object that is both adaptable and resistant, because its integrity depends not on the preservation of a few elements, but on the synergy of all the fibres that make it up.

Just as our vision of the communities of the past is influenced by our knowledge of the materials they used, if we are to imagine a different future then our ideas about technology will have to change accordingly. In my view, it is not a question of accepting or rejecting technology en bloc; those who raise the question in these terms overlook the fact that technologies are plural, and that there are infinitely many ways to relate to the materials around us better than we have traditionally done. We should always believe that a different future is possible, even if this requires us to revolutionise our perspective on how our technologies act in the world. And this means first of all rethinking the materials with which we build our lives so that they are increasingly *intelligent*, i.e. flexible and able to enter into dialogue with their environment. The idea that a material can be intelligent might appear counterintuitive; intelligence is a category we like to apply very sparingly—not least, perhaps, because it makes us feel special. But in fact, although they may be quite distinct and different from any living material, non-living materials can be far more dynamic and complex than you might imagine. After all, life itself, before coming alive, was nothing but chemistry; if there is any continuity in nature, then we must admit the possibility, if only in principle, that chemistry can produce systems of incredible intelligence and complexity.

Arachne's contest with Minerva, as depicted in the fifteenth-century manuscript *Ovide moralisée*. Image: Warburg Institute.

A Spider's Work

In the West, myths that tell of the relationship between man, or woman, and technology are often rather tormented: from the torture of Prometheus to the fall of Icarus, Greek mythology is full of warnings against the dangers of using technology to go beyond human limits, even when this is done out of necessity or for the well-being of the community. Our contemporary sensibilities, however, are often more sympathetic to the reckless figures of these myths, who appear to us as heroes and heroines in revolt against an unjust divine arrogance. Perhaps Ovid already felt this sympathy when, in the sixth book of the *Metamorphoses*, he tells the story of Arachne, a woman of humble origins but endowed with a superlative talent for weaving. According to the myth, Arachne refuses to attribute her talent to a divine gift, believing that it is exclusively the product of her own ingenuity. The goddess Athena, the patron of weaving, angered by Arachne's pride, presents herself to Arachne in the form of an old woman, recommending that she implore the goddess's forgiveness without delay. When Arachne refuses, Athena reveals herself and challenges the defiant weaver to a contest. While Athena weaves a tapestry depicting the great deeds of the Olympian gods, Arachne, with equal mastery, weaves one showing the deceptions and violence suffered by women at the hands of the gods.

At this point, an enraged Athena destroys Arachne's blasphemous tapestry and attacks her, until she is forced to commit suicide. Only at that point does Athena, merciless to the last, decide to save Arachne's life but, as punishment for her presumption, forces her to spend the rest of her days weaving her web in the body of a spider.

The myth of Arachne raises an interesting question: Do our technologies belong to us, and can we therefore use them to change what seems wrong and unjust in nature, or are they divine forces which, like the fire of Prometheus, we may borrow but must always handle with sacred reverence? Too often it is suggested that technological progress is a destiny already written, but in reality there is very little that is de-terministic in the development of the technologies we use. The path that leads to the emergence of a new technology is far from linear, and a critical view of this process is crucial to understanding technology not as a Pandora's box we have found ourselves holding, but as a canvas in which we weave one thread at a time, and where our personal and cultural perspectives carry a great deal of weight. The metamorphosis of Arachne only makes explicit an already implicit truth: that the tap-estry is inseparable from its weaver, and that technology is inextricably woven into the life of the one who constructs it.

Like weaving, the design of new materials was born in response to the practical needs of human life. This leads us to underestimate its scope, based on the assumption that an innovation that solves a practical problem cannot be worthy of theoretical consideration. For we too are convinced that the design of new materials, like weaving, is little more than 'spider's work': a boring and repetitive technical pro-cedure that has very little to teach us about ourselves or the world. We thus become entangled in a strange paradox: a great deal of scientif-ic knowledge about phenomena that are manifest only on a cosmic or subatomic scale, in times and spaces so vast or so tiny that they are intuitively elusive to our mind, end up being much better known than the principles of the applied sciences with which we build our everyday technologies. Think, for example, of the theories of relativity and quantum theory, which are undeniably fascinating and have deep epistemological implications, but which have long monopolised public discourse on science without leaving much room for anything else. It is quite likely that the average woman or man in the street will have a better grasp of the riddle of Schrödinger's cat than of the chemical structure of the polyester their socks are made from.

This reminds me of the famous story of Thales, the ancient philos-opher who, too engrossed in looking up at the stars, his face raised to the heavens, ended up falling down a well. But our well is particularly deep and dark, because if we do not find a way to act radically on our material culture, the effects, as we all know, will be irreversible. Of course, that does not mean we should stop paying attention to the more theoretical sciences, nor that we must try to make sure everyone has an encyclopaedic knowledge of every material they come into contact with. On the contrary, it means acknowledging that the ability of a science to act effectively on the world does not imply its irrelevance, but should instead prompt us to look at it more closely, because it offers us the opportunity to grasp the thread that runs back and forth between our ideas and the material world in which they are immersed. Moreover, in the face of global ecological crisis, we can no longer allow ourselves to leave unquestioned those disciplines which, well before all others, will have a direct impact upon the destiny of our species (and countless others): the applied sciences, to which we entrust the incredibly complex task of building our future. The fabric that binds Arachne to her own destiny is the same that binds ours indissolubly to the materials we choose to construct.

The divine wrath Arachne calls down on herself is perhaps linked to the fact that she is the repository of precious knowledge. It is not just her tapestry—all the infinite variety of materials that exist in nature are woven together into a kind of fabric: a multiform molecular warp that expresses the dynamic complexity of life itself. Is it possible for skilled weavers to artificially reproduce this complexity? Is it possible to imitate it or even supercede it with an artificial material? Perhaps it is no coincidence that one of the exemplary models of smart materials in nature is spider silk. The amazing properties of spider's webs, such as the ability of a very fine thread structure to support a much heavier animal and effectively trap very fast prey, were probably well known even in ancient times. And indeed the mythologies of many different civilizations contain references to spiders and their webs, often in the form of female figures associated with the art of weaving, like our Arachne. Clearly, the way in which a spider produces its own web, as well as the nature of the threads it uses to make it, are very different from those of a human weaver; and yet the incredible characteristics of spider silk are also largely determined by a kind of weaving, albeit one far more complex and finer than that of any known fabric.

It is important not only to know that spiders produce silk, but also that the spider is the only animal that can secrete and use it at every stage of its development, and is undoubtedly the animal that uses it to produce the most refined and complex structures. Furthermore, there is not just one type of spider silk: the composition of silks varies enormously from species to species, and the same spider will use between three and seven different types of silk, each produced in a particular gland in its abdomen. Moreover, although all spiders produce silk, not all of them use it for the same purposes. The vertical spider web that we immediately associate with the spider was certainly a later adaptation; the early ancestors of the spiders who colonised the emerging land-masses around four hundred million years ago used silk to protect their eggs and to reinforce their underground shelters. Even today there are still many species of spider that do not build the unmistakable spiral web familiar to us all; these are the sole province of the *Araneidae* family, which probably began to weave its vertical webs 'only' two hundred million years ago. In any case, the evolutionary pathway of spiders has been quite heavily determined by small molecular changes in their silk, which have allowed them to build increasingly functional and complex structures, ensuring their continuing success in a great many different environments.[1]

The reason why the spider's web has attracted so much attention undoubtedly lies in what we might generically define as its mechanical properties: it is a commonplace to compare the spider web's incredible strength to materials more commonly associated with mechanical strength, such as steel. One of the most popular and extravagant examples often put forward is the hypothetical ability of a silk thread as thick as a pencil to stop a fully-laden Boeing 747 in mid-flight. Actually, as intuitive as the concept of strength seems to us, this type of example only goes to show that defining the ability of a material to withstand a force is far from being a simple matter. There are rigid materials that are capable of deformation, but which, precisely because of their rigidity, are unable to withstand too much strain. There are pliable materials

---

1.   See E. Pennisi, 'Untangling Spider Biology', *Science* 358:6361 (2017), 288–91: 289.

that can undergo enormous deformation without breaking, but which would be totally unsuitable for many structural applications: no engineer would dream of building a rubber bridge. To further complicate things, many materials behave differently depending on the situation: the properties of a material can change drastically depending on the size or speed of the deformation they undergo. Another fundamental aspect of material behaviour is elasticity, which has to do with a material's ability to handle the energy it receives: an elastic material hit by a moving object is temporarily deformed, but then returns all of its kinetic energy to the bullet as it regains its original form. A material with low elasticity, on the other hand, is able to absorb and dissipate at least some of the energy it receives. In the light of these complex behaviours, the comparison of a spider web with steel—or of any material with any other—tells us little, because it does not consider their respective functions. Instead, we should understand that no material is intrinsically better or worse than another, but that each material holds a specific potential hidden within its intimate chemical structure. This is a concept with which spiders are very familiar, since they have developed a wide variety of materials with very different properties depending on their use.

The type of silk that has been the focus of most interest is 'major ampullate silk', which spiders use as the main structural element in the construction of their vertical webs. This silk is the one with the most outstanding mechanical characteristics, and toward which the greatest research efforts have been directed. The threads of major ampullate silk are considerably more resistant to tensile stress, i.e. can withstand much greater forces, than any other biomaterial: their resistance is comparable to that of high-tensile steel, although it is still lower than that of advanced synthetic materials such as Kevlar or carbon fibre. But what really makes this silk amazing is its *toughness*, which can be defined as the amount of energy a material can absorb before fracturing. Since silk is considerably more extensible than steel but with a similar tensile strength, it has a far higher toughness than almost any other known material. Another type of silk produced by spiders well known for its tenacity and resistance is the viscose silk they use to construct

A

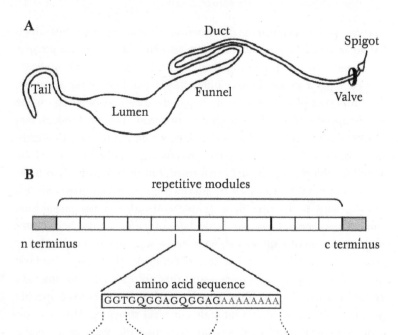

B

repetitive modules

n terminus   c terminus

amino acid sequence

| GGTGQGGAGQGGAGAAAAAAAA |

functional motif   functional motif

C

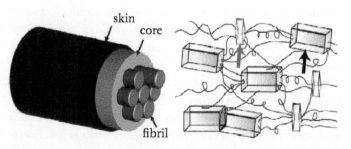

Production of ampullate silk. (A) Schematic drawing of the spider's silk glands. (B) Chemical structure of spider silk proteins, showing the repetitive amino-acid sequences. (C) Cross-section of a spider silk thread (left) and microscopic morphology of spider silk (right), showing the hierarchical structure of the material, composed of rigid and flexible components. From Lukas Eisoldt et al., 'Decoding the Secrets of Spider Silk'.

the spiral trap at the centre of their webs: this is a silk whose specific function is to catch insects that collide with it at great speed, absorbing their energy without breaking.[2]

Confronted with such characteristics, it is easy to forget that spider silk is not an innovative material designed in the laboratory, but one that has been produced for at least two hundred million years inside the belly of an organism that we usually think of as being 'low' on the evolutionary scale. What verges on the miraculous is that this material is an almost instantaneous preparation, whereas most of the synthetic fibres we use today require high temperatures, lengthy industrial processes, and polluting organic solvents for their production. But of all the incredible characteristics of spider silk, the most important is its ability to respond to stimuli in an intelligent way, adapting its behaviour and characteristics according to the environmental conditions in which it finds itself. Silk transforms from liquid to solid in a very short time, and only does so at the exact moment it is emitted from the spider's abdomen. Moreover, the mechanical response of the web changes considerably depending on the speed of the stimulus to which it is subjected, so that if the spider freefalls into it, or if an insect crashes into its threads, its resistance and toughness increase dramatically—as if, somehow, the material *knew* of the risk of an imminent fracture. This is owing to the fact that spider silk is subject to a phenomenon known as *hysteresis*, a behaviour shared by many complex systems, from materials to electronic circuits, from neurons to economic systems. Hysteresis is, in short, a memory effect: it refers to the ability of a system to undergo irreversible change when an abrupt transformation occurs, and it is one of the ways in which material structures can adapt and retain a trace of their history. When a silk thread is stretched quickly by a moving load its internal structure is reorganised so that, when the load is removed, returning to the original state yields much less energy. This energy dissipation is essentially a form of adaptation that allows the spider web to deal with abrupt environmental stimuli that would otherwise destroy it completely. On a more general level, hysteresis could be understood

2.   J.M. Gosline et al., 'The Mechanical Design of Spider Silks: From Fibroin Sequence to Mechanical Function', *Journal of Experimental Biology* 202 (1999), 3295–3303.

as one of the concepts that define our idea of what intelligence truly is. While the response of a simple mechanical system is regulated by static input-output relationships, intelligent systems do not simply yield different effects when exposed to different causes. Arguably, what defines intelligent behaviour is the capacity of a system to retain the memory of past stimuli by transmuting it into structural change.

Structure and Function

To understand the origin of these complex behaviours, we must take a closer look at the molecular structure of silk. First of all, silk is a protein, i.e. a chain—in chemistry we call it a *polymer*—of small molecular fragments called *amino acids*. In particular, major ampullate silk is obtained from a mixture of two different types of protein, which emerged at two successive moments in the spider's evolutionary history. Amino acids, the molecular units that make up proteins, are very similar to one another: every amino acid has two ends, each capable of chemically bonding with the opposite end of another amino acid, as if they were Lego bricks or puzzle pieces. Each amino acid, however, is characterised by a molecular 'pendant group' different from the others, and which gives it particular properties: in living systems there are twenty in all, which together form a sort of alphabet of life. Combining these twenty fragments yields a surprising variety of proteins, each capable of performing its own specific function, from communication between the cells of an organism to the catalysis of metabolic reactions, to specific structural functions as in the case of spider silk.

The reason why proteins are so good at doing what they do is that they have been subject to selection over millions of years of evolution. We know that amino acid sequences of proteins are encoded within genes in DNA molecules, and that evolution by natural selection acts by 'rewarding' genes that provide the best chance of survival in a given environment. In science, however, there are many different ways to interpret and answer the question *why*, and the *biological why*—the one that responds to the problem of the functioning of proteins by referring back to their evolutionary process of selection—is different from the *chemical why*. Chemistry is by its very nature a synthetic science, in the sense that, since its earliest historical origins, it has always asked how to *produce* new substances in the laboratory: for this reason, the *chemical why* is always an operational why—more like 'How is it done?' As far as proteins are concerned, what is chemically fascinating is the way in which a set of extremely simple molecular components, amino acids, can be combined to build a variety of complex objects capable of performing the most diverse functions. The *why*, in this sense, is hidden within their structure: every amino acid is capable of interacting with all the other amino acids in the same chain via its own specific side chain. The individual interactions of each amino acid are relatively simple, but when tens or hundreds of them are joined together, the interactions become incredibly complex, giving rise to specific functional structures, as when threads are woven together into fabric. This modularity, with many small simple elements interacting with one another to form complex objects, is incredibly efficient and versatile, and is something that the art of weaving and the chemistry of living organisms have in common.

The molecular fabric of major ampullate spider silk has been extensively studied, revealing a number of fascinating features. Despite the structural variability of silk from species to species, all silk proteins are characterised by the presence of two different types of amino acid sequences: some sections of the protein tend to organise themselves into rigid and strongly ordered crystalline structures, while others tend to maintain a more disorderly structure in which molecules have great freedom of movement. This alternation between rigid fragments and flexible chains is what ensures that the silk proteins will have both extremely high tensile strength and extremely high extensibility, giving the fabric its outstanding mechanical properties. However, the properties of silk are influenced not only by the amino acid sequence of the proteins that make it up, but also by the way in which individual proteins interact with one another to form a macroscopic thread. The structure of spider silk is still being studied, but it is hypothesised that the protein fibres organise themselves hierarchically into different structural levels. We can imagine the structure of silk as a series of Russian dolls: every larger structure contains a smaller one, all the way down to the sequence of amino acids of the individual proteins, and it is this stratified structure that yields spider silk's complex and intelligent response to environmental conditions.[1]

In short, the silk's astonishing ability to adapt its mechanical response to the stimuli it receives can be understood as the result of its structure: with a molecular composition that is partly rigid and partly amorphous, i.e. part crystalline and part disorderly, silk is neither a solid nor a liquid, but a material with hybrid characteristics. The presence of mobile structural elements makes it similar in some ways to a viscous liquid, capable of transferring the kinetic energy of a macroscopic mechanical stimulus—such as the impact of an insect—to its own molecules, which, moving more or less independently of one another, are able to dissipate this energy rapidly in the form of heat, thus preventing the silk from breaking, as would happen in the case of a completely rigid material. At the same time, though, the structural elements of the silk, at the moment they are subjected to tension, can organise themselves within the thread in an orderly manner, maximising their reciprocal interactions and producing an immediate reinforcing effect which makes the thread stronger and allows the spider to save itself from a sudden fall. This complex response occurs in a fraction of a second, and is not coordinated by any external 'controller'; it emerges as a result of simple chemical and physical interactions between the molecules of the material.

Similarly, the threads of spider silk also radically modify their own structure when exposed to water or ambient humidity, shortening and considerably increasing in diameter. This phenomenon is called *supercontraction* and, although its function has not yet been fully clarified, it is yet another of spider silk's fascinating properties. When water is absorbed by the silk thread, the protein chains of which it is composed fall out of alignment, returning to their ordered fibrous structure only when they dry out. As supercontraction causes the silk fibres to stiffen considerably, it is possible that this structural change is necessary to support the weight of raindrops or dew deposited on the web daily. It has also been proposed that this structural change is a kind of

1.   J.L. Yarger et al., 'Uncovering the Structure-Function Relationship in Spider Silk', *Nature Reviews Materials* 3:3 (2018), 18008.

self-repair mechanism. The microscopic structure of the silk is destroyed during the absorption of water, giving the proteins greater freedom of movement and allowing the formation of new physicochemical interactions during drying: in this way, the silk threads, subjected to the stress caused by the capture of insects, would be able to regenerate the damaged parts of their structure thanks to the morning dew. The capacity of spider silk to regenerate its structure on its own is not only an enviable ability—research into new synthetic materials capable of self-repair is an area of ongoing development—but also demonstrates once again its ability to respond intelligently to environmental stimuli.

But perhaps the most amazing aspect of silk is the mechanism by which it can transform itself from liquid to solid in a fraction of a second. This phenomenon is currently one of the most controversial, because while it is already possible to produce spider silk proteins in the laboratory using genetic engineering, transforming them into the ultra-resistant fibres used by spiders in nature would be a difficult task.[2] Contrary to what one might suppose, the glands in the spider's belly do not contain silk threads ready to be stretched into a web, but a highly concentrated aqueous solution of proteins not yet assembled into their fibrous structure. In our daily experience we are used to thinking that the transformation of a material from liquid to solid occurs by cooling, as when a molten metal wire solidifies at room temperature, or through evaporation of a solvent, as for example when we apply a varnish and let the air dry it. The solidification mechanism of spider silk, however, is unlike either of these processes. The process of fibre formation is not yet completely clear, but it is possible that it is triggered by an acidity gradient within the spider's glands. Suppose we have a dense, liquid solution to which we need only add a few drops of vinegar and instantly, threads of the most resistant material in nature emerge from the solution. How can a stimulus as simple as a change in acidity determine the formation of such a complex functional structure?

The formation mechanism of the silk is not quite as simple as this, and is probably also determined by other factors including the force

2.   A. Rising and J. Johansson, 'Toward Spinning Artificial Spider Silk', *Nature Chemical Biology* 11 (2015), 309–15.

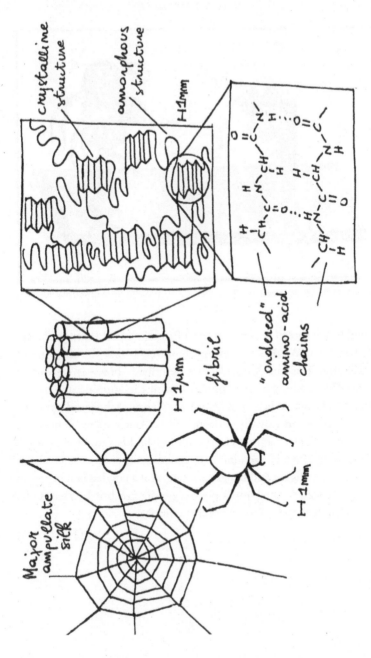

Chemical structure and hierarchical morphology of spider silk.

In the late 1800s, French missionary Jacob Paul Cambou built a hand-driven machine to extract silk from up to twenty-four spiders at a time.

applied to the thread as it is extracted from the gland through the spider's spinneret. What is interesting, however, is that the structure of the silk—the 'Russian doll' structure described above—is capable of *self-assembling* so long as it is provided with the appropriate environmental conditions. Whereas in an ordinary fabric each thread has to be woven together individually with the others, Arachne's web—once the weaver has been transformed into a spider—builds and rebuilds itself: its molecular 'threads' are able to organise themselves autonomously, creating an intelligent and dynamic weave, as if a tangled ball of cotton thread were transformed into a perfectly ordered fabric by means of a simple change in the environmental conditions.

Weaving the Future

In her book *Zeros And Ones*, the philosopher Sadie Plant examines the relationship between weaving and new technologies, especially in the context of computer science and cybernetics. She links the idea of *soft technology* to the concept of *software*, i.e. programs that are able to perform abstract processes via a material substrate. According to Plant, fabric and software are based on a similar principle:

> [T]extile images are never imposed on the surface of the cloth: their patterns are always emergent from an *active matrix*, implicit in a web which makes them *immanent to the processes from which they emerge*.[1]

In this sense, it is never possible to separate the abstract content of the software from the material process that produces it, in the same way that it is never possible to separate the pattern we see in a tapestry from the interweaving of the threads that make it up. Although the concept of software does not belong to the domain of materials science, the thought that Plant expresses here is also applicable to the *soft* material that makes up the spider's web, and to many other smart materials. The ability of a material to construct and modify its own structure autonomously suggests a form of intelligence which, unlike the centralised form we are used to thinking about, is produced continuously within the dynamic fabric of chemical and physical relations between the elements that make up the material.

If we had to choose an exemplary material, a specimen we could use to indicate the direction we ought to take when designing the materials of the future, spider silk would certainly be first in line. In addition to

---

1.   S. Plant, *Zeros And Ones: Digital Women and the New Technoculture* (London: Fourth Estate, 1998), 67 (emphasis mine).

its oft-cited mechanical resistance, it is also able to adapt to the environment, to assemble itself independently and to regenerate its structure;
it is completely biodegradable and, if we were able to discover all of its
secrets, could be produced in ordinary environmental conditions from
an aqueous solution with minimal energy expenditure. Compare this
with nylon 66, the synthetic fibre omnipresent in our clothes, which is
also a polymer made up of molecular building blocks held together by
bonds essentially identical to those that bind together the amino acids
of spider-silk proteins. However, not only are the 'building blocks' of
nylon, two molecules called hexamethylenediamine and adipic acid,
petroleum derivatives; combining them to form nylon requires a reaction that only works at a pressure of 18 atmospheres and a temperature
of 275 degrees centigrade. But energy efficiency in fibre production is
not the only reason for interest in spider webs. Several studies have
proposed using spider silk in biomedical applications as a support for
tissue regrowth; supercontraction could even be exploited to produce
artificial muscles capable of performing work in response to a simple
environmental stimulus. More generally, science has always looked to
nature as a source of inspiration for new technologies, an approach
known as *biomimicry*: the structure of a spider web, composed of different microscopic elements capable of cooperating with one another
to form a complex material, could be an inspiration for the design of
synthetic materials with equally startling properties, and with applications yet to be imagined.[2]

In a certain sense, spider history has fared better than the history of
human weavers of the past. In certain cases, the remains of one of their
ancient ancestors have been permanently trapped in amber, offering us
a privileged glimpse into a lost world; and where these finds were not
enough, evolution has left the story of their ancient art written in the
genes that code for silk proteins. The remains of Palaeolithic fabrics, on
the contrary, have been completely destroyed, and there is no genetic
archive to preserve their history for us. Human technology is far more
fragile and volatile than evolution by natural selection, subject as it is

2.   X.Y. Liu and J.-L. Li, *Soft Fibrillar Materials: Fabrication and Applications* (Weinheim:
Wiley-VCH, 2013).

to social and cultural forces that continuously influence its direction. Science journalist Leslie Brunetta and biologist Catherine Craig, in their book on the evolution of spider silk, argue that the extraordinary evolutionary success of spiders has been determined by their ability to exploit relatively small genetic mutations to obtain enormous immediate benefits in the functional properties of their silk proteins.[3]

The idea that a very small chemical change in a material can determine the survival of an entire species is an equally fascinating concept when applied to human technologies: very often, the impact that a new material has on our future extends far beyond its molecular structure, involving numerous aspects of our life and culture. Precisely for this reason, it is up to us to try to envision a world in which a material as intelligent as spider silk could play a significant role and realise its full potential. From this perspective, it may not be enough to continue thinking about technology in the same way we always have done: if when confronted with a complex and refined material such as spider silk, all we can think about is its ability to stop an aeroplane in flight, the principal limitation is obviously not technical feasibility but lack of imagination.

Contemporary materials science is always battling against this over-simplistic, hyper-optimistic view of the opportunities that new materials can afford us. The most recent and blatant example of this misunderstanding has concerned the discovery of graphene, the two-dimensional, nanometric carbon-based material that was first isolated at the University of Manchester in 2004 and has quickly become a symbol for the advancement of nanotechnology. As a researcher in the field of nanomaterials, I am often asked why graphene has not changed the world yet. In light of graphene's outstanding mechanical, electrical and thermal properties, promises of space travel, boundless energy storage and superstrong body armour have flourished in pop-science articles and books over the last decade. But fantasy tales of 'wonder materials' tell us very little of the real ways in which advancements in materials science can impact our lives and culture, and often result

3.   L. Brunetta and C.L. Craig, *Spider Silk: Evolution and 400 Million Years of Spinning, Waiting, Snagging, and Mating* (New Haven, CT: Yale University Press, 2010).

in widespread public frustration and disappointment. Nowadays, the most promising applications of graphene do not involve its use as a standalone, super-performing material, but rather focus on its integration into more complex structures, creating hybrid, composite materials with enhanced properties.[4] The behaviour of such composite materials closely resembles the mechanisms though which spider silk achieves its outstanding characteristics: dynamic self-organisation in complex material systems is gradually surpassing the need for one single, indestructible material.

Without a doubt, spider silk is one of those exceptional materials which, in materials science, is defined as *soft*: a term which includes all non-rigid, often polymeric materials with characteristics that place them somewhere on the spectrum between solid and liquid. *Morbido*, the word for 'soft' in my native Italian, suggests a more negative connotation, associating this lack of rigidity with something sick and defective: its root is the Latin word *morbus*, which means *disease* (hence the English word 'morbid'). A connotation that perhaps is not just etymological, but also tells us something about the material culture in which we are immersed. Spider silk is intelligent precisely because of its softness, which allows it to modify its structure in response to the environment. And rethinking technology in soft terms is an at once technical and cultural challenge, one that we can only address by developing an integrated vision of technology and its role in our lives.

---

4.   T. Barkan, 'Graphene: The Hype Versus Commercial Reality', *Nature Nanotechnology* 14 (2019), 904–10.

2

A MANY-HEADED THING

As a second labour Eurystheus ordered Hercules to kill the Lernaean hydra. That creature, bred in the swamp of Lerna, used to go forth into the plain and ravage both the cattle and the country. The hydra had a huge body, with nine heads, eight mortal, but the middle one immortal.

Apollodorus, *Bibliotecha*

Hydra

From the three heads of Cerberus to the snakes writing around Medusa's scalp, the idea that the seat of an organism's thoughts and consciousness might be located in more than one head has given rise to some of the most terrifying monsters in Western mythology. The most famous of these infernal creatures is undoubtedly the Lernaean Hydra, a highly poisonous sea serpent which, according to the myth, was confronted and defeated by Hercules in the second of his twelve labours. What made the Hydra a particularly fearsome adversary was not only its multiple heads—ranging from nine to fifty in different tellings of the story—but its ability to miraculously regenerate them if they were cut off, producing two new heads every time one was destroyed, thus making it impossible to kill. Hercules only managed to defeat the monster by burning the stumps of its severed heads, preventing it from continuing to multiply, and crushing the last remaining head with a boulder. In a sense the Hydra is the true nemesis of Hercules, who in turn embodies the archetype of the Western hero. His twelve labours, which involved battles with many monstrous creatures, tell a tale of the triumph of rational man over the blind forces of brute matter, a matter which boasts an almost inexhaustible capacity to produce all shapes and sizes of abominable creatures in which the bodies of animals, men, and gods are mixed to produce frightening chimeras. Hercules can only have one head, the sole seat of consciousness that makes him a human individual, while the Hydra, a polycephalous organism capable of exponentially multiplying its heads, embodies a sort of metaphysical horror: it represents the disorder and multiplicity that constantly endanger the social and cosmic order, threatening to plunge it into chaos. But I should like to strike a blow in defence of the Hydra here: this fantastic polycephalous creature is not a blind, witless monster, although it certainly embodies a form of intelligence radically different from that of its human antagonist.

Hercules's legendary struggle against the Hydra suggests the idea that precisely because it is less centralised, an organism with more heads is also necessarily less vulnerable. This is generally a valid principle when applied to any physical, biological, or social system, where a greater number of control centres guarantees greater stability in the face of environmental disturbances. Even our own organism, although endowed with a highly centralised nervous system, owes its ability to respond effectively to the environment largely to its 'polycephalous' nature—that is, its multi-cellular structure, in which various tasks are performed by specialised organs and tissues. There are however other organisms that have taken this delocalised structure to its most extreme consequences. In nature there are many animals capable of regenerating parts of their own bodies, from lizards to earthworms, but none of them have capabilities comparable to those of the legendary Lernaean Hydra. To find something that resembles the great Hydra, at least conceptually, we need to venture much further away from our particular branch of the tree of life.

In recent years a bizarre and apparently insignificant organism has begun to attract increasing scientific attention, and has even sparked widespread interest among the general public. It is *Physarum polycephalum*, also known as *mucilaginous mould* or *polycephalous slime*, a simple organism belonging to the protist realm, with an appearance similar to that of a yellowish mould, varying in size between ten centimetres and a metre in diameter. Unlike a fungus, however, *Physarum polycephalum* is capable of moving through its environment at a speed of one millimetre per second, deforming its body and forming strange 'tentacles' known as *pseudopods* which allow it to explore the world around it. Although it moves at a very slow pace for an animal, its amorphous body allows it to explore in search of the decomposing plant matter upon which it feeds. From a morphological point of view, *Physarum polycephalum* is neither a monocellular nor a multicellular organism. In fact, during the main phase of its unusual life cycle, a polycephalous slime is made up of a very large number of cells which, unlike multicellular organisms proper, are fused together into a single membrane containing the endoplasm within which the nuclei freely float; for this reason, the organism has long been classified as *acellular,* i.e. lacking real cells. *Physarum polycephalum* is not unique in kind: the group of slime moulds to which it belongs, known as *myxomycetes* (literally *muddy fungi,* even though they have no real family connection with the actual fungi) contains numerous other similar organisms, each formed of a set of cells joined together to form a single indistinct agglomeration called a *plasmodium.*

This unique cellular structure is not the only reason *Physarum polycephalum* has received so much attention. As already mentioned, its exceptional nature lies in the fact that the plasmodium of polycephalous slime exhibits behaviour similar to that of an animal: it is able to move around its environment in search of food and protection from sunlight, even though it is a far simpler organism with no tissue or nervous system. This ability to move and change shape means that polycephalous slime exhibits a set of abilities that make it seem weirdly intelligent. Where does this intelligence come from? The movements of *Physarum polycephalum* are certainly not directed by any 'control centre', because its body is little more than an indistinct, homogeneous mass of endoplasm. But then how does it orient itself in space in order to find food? How does a brainless protean jelly direct its body into coherent and organised motion? Trying to answer these questions might give us some insight into how bodies radically different from our own express other forms of intelligence, and may perhaps even allow us to make peace with the polycephalous monsters that have populated our nightmares since ancient times.

In the early years of this millennium the behaviour of *Physarum polycephalum* began to be studied in the laboratory in an attempt to find an explanation for its ability to cope with a variety of different problems. In a 2010 experiment that became famous, a team of scientists from the University of Hokkaido grew a specimen of *Physarum polycephalum* on a reproduction of the Tokyo city map, placing oat flakes—the slime's meal of choice—at the nerve centres of the city. In a very short period, the organism had managed to optimise the routes connecting all the food sources, producing a network of 'tentacles' surprisingly similar to

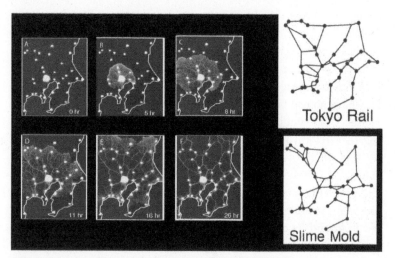

Tokyo rail network formation with *Physarum polycephalum*. From Awad et al., 'Physarum Polycephalum Intelligent Foraging Behaviour and Applications—Short Review'.

the city's rail transport network,[1] a result that is far from being obvious in advance: in computer science, the problem confronted by polyceph- alous slime here is known as the 'travelling salesman's problem', and it is know to be very difficult to solve with computational approaches. In order to optimise the most efficient route connecting n nodes of a network, it is necessary to individually evaluate all possible routes, the number of which multiplies exponentially as n increases, with a conse- quent exponential increase in the calculation time required by the algo- rithm. The fact that an extremely simple organism was able to find an efficient solution to such a complex problem immediately attracted the attention of scientists from a variety of different disciplinary fields. In- formation scientist Andrew Adamatzky, director of the Unconventional Computing Laboratory at the University of the West of England, has extensively studied the behaviour of *Physarum polycephalum*, extend- ing the Tokyo experiment to a variety of other geographical territories and in each case achieving very accurate reproductions of manmade

1.   A. Tero et al., 'Rules for Biologically Inspired Adaptive Network Design', *Science* 327:5964 (2010), 439–42.

transport networks. In one of the most curious experiments, a specimen of polycephalous slime was even placed on three-dimensional models of the Moon and Mars to evaluate possible approaches to the colonisation of extraterrestrial planets.[2]

From this point on, a whole subfield of academic studies developed around the abilities of polycephalous slime to solve complex real-world problems, with Adamatzky hailed as the pioneer of a discipline known as 'Physarum computing'. The proposal is to use these living systems as real computational machines, exploiting their ability to optimise efficient pathways between different food sources.[3] These types of optimisation problems are particularly difficult to solve with conventional approaches, but occur often in our daily lives: not only in road planning but also, for example, in calculating the most efficient routes for city traffic. Even though polycephalous slime is very different from a conventional computer, it can do something that our computational machines are not capable of: it can be used to perform what in computer science is known as *morphological computation*, i.e. it is able to 'think with form', modifying its body to build complex networks that would require a prohibitive amount of calculation time for ordinary computation.

Physarum polycephalum goes in search of food by expanding almost homogeneously into the surrounding environment, then retracting the parts of its body that are useless for providing nourishment, thus forming a network of little yellowish tubes that look like some kind of bizarre circulatory system. When it retracts from a surface, the slime deposits a substance that signals to its body not to expand in that particular region again, thus constructing a genuine *spatial memory* of its environment. This ability to remember what surrounds it, albeit in an unconventional way, combined with its ability to control the flow of its own endoplasm by exploiting a protein structure similar to that of our own muscles, allows the polycephalous slime to colonise its

---

2.   A. Adamatzky et al., 'Physarum Imitates Exploration and Colonisation of Planets', in A. Adamatzky (ed.), *Advances in Physarum Machines: Sensing and Computing with Slime Mould* (Basel: Springer, 2016), 395–410.

3.   Adamatzky (ed.), *Advances in Physarum Machines*.

environment with an efficiency comparable to that of human beings, even without a brain and, as far as we know, without being aware of it. Since *Physarum* has no eyes or sense organs, it cannot coordinate its movement on the basis of a *representation* of the reality around it. Its intelligent behaviour emerges from a multitude of simple biochemical mechanisms acting locally in every part of its body.

Clearly, no one thinks that Physarum computing will in the near future replace the computational machines we are used to. The use of a living system as a medium for computation has obvious limitations, first of all the need to keep it alive and to reprogram it every time the need to perform a new operation arises. Even if there might be particular technological applications for the use of polycephalous slime as a problem-solving system, these studies, rather than aiming to develop a real technology, are more like a sort of thought experiment, making it possible to rethink computation and intelligence in new ways. Whereas in a traditional computer all information is processed by a *central processing unit* that performs a series of complex operations one at a time, one of the attractions of the possibility of using alternative materials for computation lies in their ability to perform a myriad of simple operations in parallel, yielding results which, in some cases, can far exceed those possible with conventional computers.

Intelligent Jelly

'Polycephalous' slime is by definition a many-headed creature, but perhaps it would be more accurate to say that it has no head at all. In other words, it does not have one or more central control systems capable of collecting information from the environment, 'calculating' an appropriate response, and then 'transmitting' to the rest of the body the impulses needed to produce a coordinated movement in the direction of food. In this sense, its intelligence is certainly very different from that of a mammal or a computer: its complex response to the environment is not processed through a single organ, but is the result of a set of numerous microscopic phenomena that occur on a molecular scale in all parts of its body. If we look closely enough at the movement of polycephalous slime, we can see that the membrane that divides it from the outside world is covered with receptors, tiny molecular sensors capable of binding to a specific chemical substance present in the environment. These receptors initiate a chain reaction that triggers a transformation in the protein structure of the organism, which then modifies the volume and viscosity of the cellular endoplasm, causing it to expand or contract.[1] No centralised coordination is involved in this mechanism: it is the continuous and delocalised dialogue between the organism and its environment that makes the movement emerge. The intelligence of *Physarum polycephalum* is built into the interface: its brain, if we can call it that, is precisely its surface, the cell membrane that both separates it from the world around it and allows it to actively interact with its environment.

Physarum polycephalum is still a living organism, albeit an unusual one. However, there are non-living artificial materials that exhibit

---

1.   B. Álvarez-González et al., 'Cytoskeletal Mechanics Regulating Amoeboid Cell Locomotion', *Applied Mechanics Reviews* 66:5 (2014), 0508041–05080414.

very similar behaviour to that of polycephalous slime: without any centralised control, they are able to move independently and shape themselves in response to many different stimuli. Although the chemistry of living organisms is incredibly complicated, the idea behind the intelligent movement of polycephalous slime can be broken down into two relatively simple elements: the ability to perceive an environmental signal, which may be chemical, as in the case of the search for food, or of some other nature, e.g. luminous, thermal, magnetic, or electrical; and a dynamic structure capable of inducing a macroscopic change in volume, shape, or physical state in response to the stimulus received.[2] These two elements must be integrated in a more or less direct way, so that the macroscopic response manifests itself every time the corresponding stimulus is present. Materials capable of responding to environmental stimuli in this way are commonly called *smart materials*, and include a growing variety of chemical systems capable of responding to various different signals.

Defining a smart material with any precision is not so easy, because all the materials around us are to a certain extent capable of changing their appearance and properties depending on the conditions in which they find themselves: for example, an ice cube melts as soon as the ambient temperature gets above zero degrees, but no one would define water as a smart material capable of responding to thermal stimuli. A lump of coal burns in the presence of oxygen, but it is clear that this response is qualitatively different from that of an organism that moves in search of food in response to a chemical stimulus. In a smart material, the environmental response must be controlled and reversible, and must not result in a complete destruction of the structure of the original material.

The first class of intelligent materials to emerge was that of *shape-memory alloys*, discovered in 1932 by the Swedish scientist Arne Ölander, who observed that an alloy of gold and cadmium was capable of exhibiting unusual, 'almost elastic' behaviour. But it was not until the 1960s that the potential of these materials began to be realised, in

2.   M. Bengisu and M. Ferrara (eds.), *Materials that Move: Smart Materials, Intelligent Design* (Basel: Springer, 2018).

part thanks to the discovery of numerous other alloys with similar characteristics. Shape-memory alloys can be 'trained' to remember a precise shape assumed at high temperature, so that if they undergo deformation at room temperature, they can be returned to their original shape simply by heating their material. Many crystalline solids change their atomic organisation when the temperature varies, taking on different microscopic structures, also called *phases*, when heated or cooled. These phase shifts are similar to those with which we are all familiar, such as freezing or evaporation. Unlike what happens when we melt an ice cube, however, they do not completely destroy the structure of the solid, but merely change the position of its atoms with respect to one another. In the case of shape-memory alloys, the high-temperature phase, called *austenite*, is not deformable, while in the low-temperature phase, known as *martenite*, atoms can move easily relative to one another and the material can therefore be deformed. Any deformation undergone by the material while at a low temperature, in the martenite phase, is eliminated when the alloy is heated and returns to the austenite phase: for this reason, each time it is heated, the alloy always returns to the shape it was given at high temperature, regardless of any deformations it may have undergone at room temperature. Shape-memory alloys may have applications in aeronautics, where structural components capable of reversible deformations could replace the heavy hydraulic actuators currently used to manoeuvre aircraft ailerons. However, while the ability of these alloys to 'remember' their shape is fascinating, application of these systems is subject to many limitations, because their behaviour is restricted to a very specific class of metallic materials that only respond to an equally specific environmental stimulus.

Things start to get more interesting when we enter the world of polymeric materials, which are incredibly versatile, flexible, and exhibit a wide variety of chemical and structural traits. Indeed, the mechanisms of environmental response typical of all living systems are implemented through proteins, which are just a special type of polymer. The potential applications for polymers capable of responding to environmental stimuli is endless, but one of the most innovative areas is that of *soft robots*: automata completely devoid of hard parts, whose

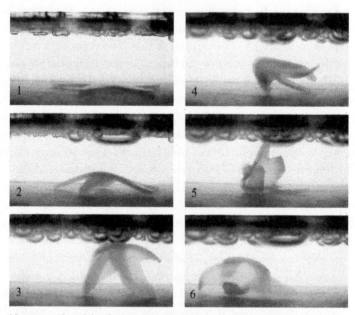

Movement of starfish gel robot. From M. Otake et al., 'Motion Design of a Star-fish-shaped Gel Robot Made of Electro-active Polymer Gel', *Robotics. Aut. Syst.* 40 (2002), 185–91. Image: Courtesy the author.

appearance and behaviour recalls that of molluscs, and which could be used to perform delicate and precise tasks, for example surgical operations that would be unfeasible using traditional robots. In 2016, researchers at Harvard University announced the creation of the first autonomous, fully soft robot.[3] The robot, named Octobot, was 3D printed in the form of an octopus by using a hybrid polymeric nanomaterial and it was equipped with an internal microfluidic system propelled by a mixture of hydrogen peroxide and platinum that autonomously controlled the robot's movements. Although entirely composed of soft materials, the design of Octobot still relied on a centralised control unit that recalled the architecture of traditional automata. Progress in materials science, however, has gradually unlocked the possibility of doing without such centralised control systems, by using materials

3.   M. Wehner et al., 'An Integrated Design and Fabrication Strategy for Entirely Soft, Autonomous Robots', *Nature*, 536 (2016), 451–55

which, not unlike *Physarum polycephalum*, are able to 'feel' their environment with every part of their bodies. One of the first notable examples of such attempts dates back to 2002 when, using a hydrogel, i.e. a polymer 'jelly' made up mostly of water, a team of researchers at the University of Tokyo designed a *soft robot* in the shape of a starfish with a striking resemblance to the real mollusc, capable of deforming itself and moving through an aquatic environment in response to an external electric field.[4] Recently, in 2021, a group of scientists at the University of California have used a new smart hydrogel material to produce yet another generation of octopus-shaped robot.[5] This material is not only capable of moving in reaction to both light and thermal stimuli, but is also self-sensing, meaning that it can produce a measurable electrical response to its own movements.

Hydrogels like this can be modified in a number of ways to be sensitive to countless different signals. In addition to the response to electrical stimuli, as in the case of the starfish, soft robots can also be designed to react specifically to certain chemicals, in the same way that polycephalous slime is sensitive to the chemical signals released by food in its immediate vicinity.[6] Different systems capable of expanding and contracting in the presence of different substances have been studied; very often this phenomenon is triggered by 'molecular switches' contained within the polymer structure which can promote or inhibit interaction between the polymer chains, leading to an overall contraction or dilation of the material. Using this type of mechanism has made it possible, for example, to design a hydrogel capable of releasing insulin directly into the bloodstream selectively, i.e. only when the system detects a high glucose concentration.[7] In an article published in *Science*, Professor Jian Ping Gong of Hokkaido University and her collaborators

4. M. Otake et al., 'Motion Design of a Starfish-shaped Gel Robot Made of Electro-Active Polymer Gel', *Robotics and Autonomous Systems* 40:2–3 (2002), 185–91.

5. C.-Y. Lo et al., 'Highly Stretchable Self-Sensing Actuator Based on Conductive Photothermally-Responsive Hydrogel', *Materials Today* 50 (November 2021).

6. H.-J. Schneider (ed.), *Chemoresponsive Materials. Stimulation by Chemical and Biological Signals* (Cambridge: Royal Society of Chemistry, 2015).

7. A. Matsumoto et al., 'Synthetic "Smart Gel" Provides Glucose-responsive Insulin Delivery in Diabetic Mice', *Science Advances* 3:11 (2017), eaaq0723.

reported that they had successfully synthesised a hydrogel capable of increasing its mass and strength after being subjected to repeated mechanical stress, similarly to how our own muscles can be trained.[8]

These examples show how the use of intelligent materials may pave the way for the design of automata increasingly similar to living organisms, combining sensitivity with flexibility and adaptability. Rather than anthropomorphic tin cans with a computer installed in their heads—the classic image that still springs to mind when we think about robotics—the robots of the future may be built with soft water-based materials far more similar to our own flesh than we had anticipated. The idea that intelligent technology should be able to receive signals from the world around it now seems quite obvious to us, but it has always been taken for granted that this interaction would be channelled through electronic 'sense organs'—video cameras and microphones—capable of mimicking human sight and hearing, the two senses to which, as a species, we have always attached the greatest importance. The truth is that in many cases the automata to come will not resemble us at all: they will be amorphous and gelatinous like amoeba, or they will look like bizarre invertebrates, capable of perceiving a world of signals that are completely inaccessible to us; yet they will be complex integrated systems, equipped with a body and capable of 'feeling' with every inch of the materials that make them up.

---

8.   T. Matsuda et al., 'Mechanoresponsive Self-Growing Hydrogels Inspired by Muscle Training', *Science* 363:6426 (2019), 504–8.

Under the Skin

When attributed to materials, the term 'intelligent' normally describes a set of behaviours including the ability to respond to certain stimuli, without implying that a material is intelligent in the same way as a person or animal. Most of the research in the field of intelligent materials focuses on very particular application-oriented materials, since the possibility of using a material capable of 'feeling' its surroundings is potentially useful in many different contexts, from engineering to medicine. Usually an intelligent material is designed to be sensitive to one or two stimuli at a time, whereas a living body, however simple, is capable of perceiving a vast number of different signals from its environment. In the context of materials science, no concerted effort has been made toward the construction of increasingly sentient automata; rather, efforts go toward designing materials that are capable of performing certain determinate functions more and more efficiently. In this sense, the intelligence of materials is better understood as a side effect of the need to optimise the functionality of our technologies, rather than as a specific project to 'animate' non-living matter. And yet, nothing prevents us from imagining the construction of a arbitrarily complex material capable of responding to many stimuli simultaneously.

In reality, the idea of designing artificial materials which, like us, are able to 'feel' the world around them is nothing new. One of the most significant advances in this direction is an advanced device known as *e-skin*, an 'electronic skin' that could be used to provide robots with a touch sensitivity similar to that of human skin. Tactile experience, after all, cannot be reduced to a single stimulus, but consists of a variety of different sensations that together add up to an integrated experience of the world. In a recent article published in *Science*, researchers described the preparation of an artificial skin built from a polymeric matrix incorporating nanoparticles of silver. This material, like our skin, is sensitive to temperature, pressure and humidity, is flexible and adaptable to uneven surfaces, is able to repair itself without loss of functionality, and is also completely recyclable.[1] The possibility of constructing such materials implies that at least some aspects of our experience in the world can be transferred into artificial bodies completely different from our own; however, there is understandable resistance to attributing what is perceived as a fundamental human experience to the rest of the matter around us.

The possibility of attributing intelligence to non-human subjects has always been a controversial issue. Our foremost encounter with other intelligences is doubtless the meeting of human and animal: although many of the animals with whom we share our space and our lives are clearly capable of feeling and understanding the reality around them, we find it difficult to accept that they may have an inner experience similar to ours. This kind of resistance functions to perpetuate dynamics of exploitation and violence with which our species is all too familiar and which, perhaps, would not be possible in the light of a recognition of the other's inner experience. The problem becomes a little more complex when we not only have to recognise our experience of the world in someone else's body, but are confronted by a form of intelligence that has very little to do with our own. The philosopher of science Peter Godfrey-Smith deals with this problem in his essay *Other Minds*, setting out from the encounter with octopi and cuttlefish, creatures endowed with a highly developed nervous system that has been selected following an evolutionary path completely independent of our own.[2] What is most interesting, according to Godfrey-Smith, is not so much the observation that cuttlefish and octopi are intelligent—that they are just as capable as many mammals of solving complex problems—but rather the fact that these animals are endowed with a form of intelligence that is qualitatively very different from human intelligence. The nervous

1.   Z. Zou et al. 'Rehealable, Fully Recyclable, and Malleable Electronic Skin Enabled by Dynamic Covalent Thermoset Nanocomposite', *Science Advances* 4:2 (2018), eaaq0508.
2.   P. Godfrey-Smith, *Other Minds: The Octopus and the Evolution of Intelligent Life* (London: Collins, 2016).

system of cephalopods, in fact, is delocalised to the point that these creatures are capable of thinking not only with their brains, but also with their arms. And that's not all: their bodies are covered with receptors that allow these animals to 'see with their skin' and 'taste' what they touch with their suckers—experiences so alien to us that they are completely incomprehensible. Precisely because of their decentralised intelligence and soft bodies, cephalopods have been a recurring aesthetic and technical paradigm in our imagination of the technologies of the future. In 2019, collaborative artist Orphan Drift created a video installation titled *If AI Were Cephalophod*, where they foreshadowed a future artificial intelligence based on the octopus's mind, questioning the way in which our anthropocentric presumptions about intelligence shape our contemporary technologies.

> If AI were cephalopod it would have three hearts and see with its skin. If AI were cephalopod it would haunt the human imagination as monster, weaving its way into legends and mythologies. If AI were cephalopod it would be 500 million years old. [...] If AI were cephalopod we would never presume to fully understand it.[3]

As evolutionarily distant as they are from us, cephalopods are nonetheless endowed, just like humans, with a nervous system. As already observed in the case of polycephalous slime, however, a nervous system is not a prerequisite for the manifestation of intelligent behaviour; on the contrary, the construction of smart materials seems to demonstrate that intelligence can be expressed, at least in some form, even in a non-living body. What lies at the heart of the problem, then, is the question of whether a system's ability to *behave intelligently* is a sufficient condition to define it as, in fact, *intelligent*. Indeed this is precisely one of the strongest objections that can be made to the intelligence of materials: the fact that a material is able to respond to environmental stimuli similarly to an organism with a brain does not necessarily mean that material can be considered intelligent in the strict sense.

---

3.   Orphan Drift (Ranu Mukherjee and Maggie Roberts), 'If AI Were Cephalopod', multi-channel video installation, San Francisco, 2019.

The elephant in the room, when it comes to nonhuman and nonliving intelligences, is naturally artificial intelligence, to which similar objections have often been made. Although they are distinct problems, the debate that has developed around the possibility of programming intelligence into a computer can also be applied, *mutatis mutandis*, to the case of intelligent materials. The philosopher John Searle, in a famous 1980 article entitled 'Minds, Brains and Programs', develops a harsh critique of the possibility of what he calls 'strong AI', i.e. an artificial intelligence capable not only of simulating intelligent behaviour, but of actually thinking.[4] To corroborate his argument, Searle proposes a thought experiment that has become famous as the *Chinese Room Argument*: imagine that a man with no knowledge of the Chinese language is locked in a room, where pieces of paper containing messages written in Chinese are handed to him. His task is to choose the correct answer from a selection of available answers. To help him in this task, the man in the room also has an 'instruction manual' written in a language he knows, in which the rules for associating each message with the corresponding answer are described in detail. To the eyes of an outside observer, unaware of what is happening in the room, the man may seem perfectly capable of understanding Chinese; however, it is clear that he does not actually have any real understanding of the meaning of the incoming and outgoing messages. According to Searle, artificial intelligences act in the same way, simulating intelligent behaviour based on a set of predefined rules, but without having any real comprehension of what they are doing.

Having presented his thought experiment, Searle goes on to rebut some possible objections to the his argument. One of the most interesting replies to the Chinese Room Argument is that, although the man inside the room is not actually capable of understanding Chinese, *the system as a whole* made up of the room, the man, and the material available to him, is indeed capable of understanding. According to Searle, though, it is absurd to think that 'while a person doesn't understand Chinese, somehow the conjunction of that person and bits of paper

---

4.   J. Searle, 'Minds, Brains and Programs', *Behavioral and Brain Sciences* 3:3 (1980), 417–57.

might understand Chinese'. If, like Searle, we accept that the human brain itself is actually capable of real understanding, we must also recognise that a single neuron, although part of the 'brain system', is not in itself capable of thinking or understanding anything. It is only by uniting many neurons into a larger system connected to itself and to the world—a system that includes not only the brain, but the body and the environment as a whole—that understanding can emerge.

Being in the World

The main limitation of Searle's thought experiment is that it lays bare a widespread but essentially unfounded prejudice about the nature of intelligence. His thesis is based on a model that could be called 'the man in the control room': intelligence is interpreted as the result of a unitary already-developed consciousness located somewhere inside the brain, which functions by receiving a series of inputs from outside, understanding them and then processing an output in response to them. This heavily centralised view of cognition probably derives from the self-reflexive tendency of our inner life, by virtue of which our consciousness acts as a mirror, reflecting the world around it, while at the same time continuously reflecting itself. This 'representational' cognitive model implies that intelligence is to be identified with a centralised model of consciousness: the only authentic form of cognition would be one that builds a model of reality before being able to act upon that reality. On the contrary, for an organism like polycephalous slime or an intelligent synthetic material, there is no representation of reality that precedes and directs action. Instead, intelligence and action are one and the same: every signal that comes from outside determines an immediate and contemporary response to the stimulus received. This shows that self-reflection is not a necessary condition for having an experience of the world; and after all, there are many occasions even in our own lives when our self-reflexive mind is silenced and we are immersed in a sort of 'pure experience' of reality—something that is actively sought after in some forms of meditation. In reality, there is no homunculus inside our head doing all the understanding, and there is no room, real or metaphorical, in which our knowledge of the world is contained.

Searle himself seems to concede this when, in laying out the constructive part of his argument, he argues that it is conceptually possible to build an intelligent machine. According to Searle, the problem with artificial intelligence as commonly understood is that it insists on interpreting intelligence as a *program* that can be implemented and operated independently of the bodily substrate in which it is realised, whereas the bodily substrate is precisely what makes thought possible. In other words, for Searle intelligence is exclusively a *hardware* problem: artificial intelligence may well be able to *simulate* intelligent behaviour with a pile of silicon, nuts and bolts, but this simulation will have nothing to do with intelligence itself. However, Searle does argue that, were we able to build machines that were truly brain-like, especially from a material point of view, then it would in principle be possible to build an artificial intelligence truly worthy of the name. In Searle's own words:

> I see no reason in principle why we couldn't give a machine the capacity to understand English or Chinese, since in an important sense our bodies with our brains are precisely such machines. But I do see very strong arguments for saying that we could not give such a thing to a machine where the operation of the machine is defined solely in terms of computational processes over formally defined elements; that is, where the operation of the machine is defined as an instantiation of a computer program. It is not because I am the instantiation of a computer program that I am able to understand English and have other forms of intentionality (I am, I suppose, the instantiation of any number of computer programs), but as far as we know it is because I am a certain sort of organism with a certain biological (i.e. chemical and physical) structure, and this structure, under certain conditions, is causally capable of producing perception, action, understanding, learning, and other intentional phenomena. And part of the point of the present argument is that only something that had those causal powers could have that intentionality. Perhaps other physical and chemical processes could produce exactly these effects; perhaps, for example, Martians also have intentionality but their brains are made of different stuff.[1]

---

1.   Ibid., 422.

In putting forward these considerations, Searle is making a very strong statement about the nature of intelligence: he is arguing that every mind is necessarily a material phenomenon that emerges from a set of physical and chemical elements placed in relation to one another—a potentially revolutionary idea that Searle does not seem to be able to extend beyond the conventional anthropomorphic conceptions of intelligence, since he continues to merely think about 'brains made of different material' and stops short of the possibility that a brain, understood as the 'command centre' of the body, is only one of many ways in which thought may come to be organised within matter.

In reality, not only the brain, but bodies as a whole, along with the world around them, intertwine and actively contribute to cognition. Indeed, the discovery of the intelligence of certain materials seems to point in this direction: smart materials are also in a sense AIs, but AIs whose functioning is inseparable from the material of which they are composed. When *Physarum polycephalum* secretes a substance that acts as a trace of its past, preventing it from re-expanding onto surfaces it has already explored, it is not simply receiving and processing a stimulus from the outside world, it is actively constructing the conditions of its own experience, defining its own space-time, and extending its mind into its surroundings. This relationship between body, mind, and environment is perhaps the most significant aspect of the intelligence of materials: precisely because the mind is no longer confined inside a brain, but expresses itself in a body that communicates with other bodies, intelligence becomes a relational affair, manifesting itself in a material network that transcends the boundaries of any individual. This interpenetration of thought and world is expressed in the concept of the *extended mind*, introduced for the first time by Andy Clark and David Chalmers in a 1998 article where they argue that consciousness and cognition are two distinct phenomena: while the former must necessarily take place within the brain, the latter can also be outsourced and delegated to other material structures which, although not necessarily conscious, actively fulfil a cognitive function.[2] This happens all

---

2. A. Clark and D.J. Chalmers, 'The Extended Mind', *Analysis* 58:1 (1998), 7–19.

the time in our daily lives: whenever we make a note in a notebook or diary, whenever we use an electronic calendar or communicate with another person via smartphone, we make use of a physical extension of our mind. The most primitive form of extended mind is our own hands, which we have always used to count, and whose 'digital' structure has definitely determined the decimal arithmetic we have learned since we were children. Faced with the evidence that all of our mathematical thought has developed around an organ that lies outside our brain, it is difficult to deny that our body as a whole is an integral and indispensable part of our mind.

The concept of the extended mind suggests something fundamental about the nature of cognition: that it is not possible to separate a mind from the world in which it is immersed. Rather than seeing it as an intrinsic feature of the brain, then, we can look again at intelligence as a widespread and decentralised phenomenon which emerges from the way in which different bodies—human and non-human, living and non-living—relate to one another. This idea fits into the broader context of *embodied cognition*, a theory of the mind that challenges the clear separation between mind and body, and finds its roots in the thought of the French philosopher Maurice Merleau-Ponty. For Merleau-Ponty, embodied experience is not simply the effect of self-reflection—the Cartesian *cogito*—but emerges as an integrated result of body perception. Perception, rather than being a unidirectional cognitive movement directed at an object 'out there', is constructed and modified in relation to the world, which in turn is determined by the perceptual capacities of the body. 'To be a consciousness, or rather *to be an experience*, is to have an inner communication with the world, the body, and others, to be with them rather than beside them'[3]—which, according to Merleau-Ponty, means *to be in the world*, an expression that indicates a condition of openness and reciprocal determination between the subject and its environment, a condition of pure experience that precedes representation. This way of describing experience lends itself to being extended to nonhuman minds as well, because it does not depend

3.   M. Merleau-Ponty, *Phenomenology of Perception*, tr. D.A. Landes (London and New York: Routledge, 2010), 99.

strictly on self-consciousness, but only on the existence of a body in a continuous relation with its own world.

> The gaze obtains more or less from things according to the manner in which it interrogates them, in which it glances over them or rests upon them. Learning to see colors is the acquisition of a certain style of vision, a new use of one's own body; it is to enrich and to reorganize the body schema. As a system of motor powers or perceptual powers, our body is not an object for an 'I think': it is a totality of lived significations that moves toward its equilibrium. Occasionally a new knot of significations is formed: our previous movements are integrated into a new motor entity, the first visual givens are integrated into a new sensorial entity, and our natural powers suddenly merge with a richer signification that was, up until that point, merely implied in our perceptual or practical field or that was merely anticipated in our experience through a certain lack, and whose advent suddenly reorganizes our equilibrium and fulfills our blind expectation.[4]

4.   Ibid., 154–55.

Through the Looking-Glass

The embodied cognition perspective provides us with valuable tools with which to confront the strange forms of intelligence that animate matter in its various guises. Although it is not possible for us to access the experience of the reaction of a sensitive gel to a certain chemical signal, just as it is not possible for us to access the experience of the octopus 'seeing with its skin', we cannot fail to recognise that these systems are bodies capable of some form of thought, because they are able to modify their own structure in order to act within the world. This, of course, does not mean that a material is endowed with consciousness, but that consciousness as we know it, although it appears to us to be a fundamental aspect of our experience, is not strictly necessary for the production of thought. Every intelligent material defines and constructs its own world, made up of signals and experiences that may be very different from ours, and with which it establishes a dynamic and continuous relationship. The recognition and study of these forms of experience is the gateway to a radical extension of our human perspective: working together with the materials we design and use, we can open up to a multiplicity of other perspectives that can then be integrated into a new technological, social, and political understanding of the world.

The theory of embodied cognition is especially useful for disrupting the conception of the human mind as an entity separate and independent from the body that produces it, by highlighting the physical, chemical, and social processes that make our thinking possible. However, limiting the potential of this theory to the human being is reductive, because this vision of body and mind allows us not only to imagine, but also to *design* new technologies which, in the future, could be integrated with our bodies, if they are not already. In this context, it is also essential to understand that this perspective on technology, which we might call *posthuman*, is not independent of the materials we use to put it into practice, just as human thought, in its apparent separation from the body, is in fact deeply rooted in the biochemical matter of which we are made. Here the concept of *softness* once again comes to our aid: the most innovative intelligent materials are all characterised by a dynamic and flexible structure, which contrasts with our traditional conception of robotics and of technology in general. Soft robots are able to *perceive a world* precisely because their entire structure is involved and transformed in the act of perception, in the same way that a spider silk thread reacts to a sudden impact by rearranging its own internal architecture. But the category of *softness* applies not only at the strictly chemical-physical of a material; at a more abstract level, it also indicates a specific way in which several parts of a whole can work together to build a mind.

Andy Clark develops this idea in his essay *Being There*, in which he presents in detail the theories of embodied cognition and their possible technological consequences.[1] In his book Clark introduces the distinction between *hard assembly* and *soft assembly*, two opposing approaches to the construction of intelligent systems. In hard-assembled systems, which include most conventional machines, centralised control is exercised over every action of the system, which requires an exact evaluation of the environmental parameters and the position of every single part of the organism. On the contrary, a soft-assembled system is based on a delocalisation and diffusion of control: each part of the organism can act locally in a simple and immediate way, while the overall intelligent behaviour arises as an *emergent property* of a community. Soft assembly guarantees greater robustness, i.e. it allows the system to adapt to a greater variety of conditions because it allows the environment to actively participate in the definition of the behaviour. Clark writes:

> In sum, the task is to learn how to soft-assemble adaptive behaviors
> in ways that respond to local context and exploit intrinsic dynamics.

---

1.   A. Clark, *Being There: Putting Brain, Body and World Together Again* (Cambridge, MA: MIT Press, 1998).

Mind, body, and world thus emerge as equal partners in the construction of robust, flexible behaviors.[2]

As biology shows us, the best way to produce intelligent soft-assembled systems is to make then from soft materials, i.e. materials that are already intrinsically equipped with only partially ordered repetitive structures and are capable of self-organising in space—for example, proteins or synthetic polymers. These materials make the execution of complex operations much easier than materials with rigid and predetermined structures, because they are able to perform at least some of the processes involved in perception and thought by exploiting the dynamic evolution of their own structure. Like *Physarum polycephalum*, these materials are also capable of 'thinking with form', modifying their internal organisation to process the information they receive from the outside world. From a more conceptual point of view, these technological approaches demand that we radically alter our perspective on the world 'out there'. We are used to thinking of our perceptual experience as a mirror in which we see the reflected image of an objective reality always separate from us. It is not really important whether we believe that this reflection is perfectly accurate, skewed, or faulty in some way: in any case, the perceived object does not actively participate in cognition. Perceptual processes, however, as the theory of embodied cognition teaches us, are not simple reflections, but involve a continuous exchange between subject and world. Intelligent materials, whether synthetic or natural, are open systems continuously traversed by the environment. In this sense, they can act as magical portals, allowing us to step through the looking-glass, breaking with the idea of human intelligence as a simple reflection of an objective reality and opening our eyes to a universe far more sentient and far more dynamic than we could imagine. This requires us, first and foremost, to reconsider our relationship with the world and with the other minds with whom we share our lives, whether living or not, whether natural or artificial. A world populated with objects is a world in which relations

---

2.   Ibid., 45.

of domination can easily be established, because it allows us to deny the subjective experience of the other; but this denial also has the effect of irremediably mutilating our perception of reality, locking us in the dead-end closed room of our anthropocentric perspective. A more open conception of subjectivity requires us to think of technologies not as mere tools, but as extensions of our mind that allow us to expand our experience of the world into unknown territories. In the process, the materials that we design are hybridised with and continuously develop along with our body and our culture, blurring the boundaries between matter and mind.

In an article entitled *Nature's Queer Performativity*, the philosopher and physicist Karen Barad considers a number of natural systems, from amoebas to lightning, which highlight the deeply relational and delocalised nature of the subjective.[3] According to Barad, our vision of a world made up of rigid and independent units that interact with one another is a simplification of a more complex reality in which the relationship between subject and object is fluid and is constructed at the moment of mutual encounter. To designate this relationship Barad coins the term *intra-action*, contrasting it with *interaction* by highlighting the idea that two physical objects that come into contact do not merely 'collide', but interpenetrate and define one another. This nuanced view of the boundaries of subjectivity is not only a scientific fact confirmed by the many natural phenomena that cannot be explained in terms of a traditional conception of identity, it is also a political and cultural manifesto that invites us to rethink matter in new terms. It is in this sense that Barad uses the word *queer* to define all material phenomena, natural or cultural, that contribute to deconstructing rigid categories of identity through an openness of the subject-object relationship. What we are often tempted to define as something 'against nature'—such as the headless body of polycephalous slime, made up of an amorphous aggregate of innumerable individuals merged together—is actually capable of revealing to us the authentic face of our own nature and that of the world around us.

---

3.   K. Barad, 'Nature's Queer Performativity', *Qui Parle* 19:2 (2011), 121–58.

3

THE PATTERN
WHICH CONNECTS

Then it is that a ghostly legend wakes to new life in the hidden recesses of my mind, the legend of the Golem, that man-made being that long ago a rabbi versed in the lore of the Cabbala formed from elemental matter and invested with mindless, automatic life by placing a magic formula behind its teeth.

Gustav Meyrink, *The Golem*

Golem

The way in which a set of simple objects interact to form a complex structure is one of the most fascinating aspects of the materials around us. As human beings, we have always been able to assemble material components to build artificial objects capable of performing different functions; our ability to build machines and structures is astonishing, and yet the way we transform the material around us is in no way comparable with the assembly processes of nature, in particular those that are characteristic of living matter. How does a plant grow? How is a diamond formed? Why do we have to work so hard to design, build, and repair our machines while the organisms that surround us, not to mention our own bodies, are capable of doing so unconsciously and totally autonomously? The distance separating our technologies from natural organisms seems like an unbridgeable abyss, dividing us from a lost paradise of infinite abundance and eternal regeneration.

The key word here is *spontaneity*: what strikes us when watch a living organism grow, or observe the perfect symmetry of a crystal, is not so much its shape or its function as the way in which these macroscopic structures emerge autonomously, *bottom up*, making use of chemical ingredients present in the environment to give rise to a higher order. In contrast, most artificial structures are constructed by humans using what we might call a vertical approach, in which a pre-constructed, external plan directs the assembly of individual material components *top down* in order to produce a given object. In natural processes of assembly there is no such plan, or, if there is, it is one completely internal to the physical system that is its product. Spontaneity may be an intuitive concept, but it is far from trivial from a physical point of view. In everyday language, spontaneity suggests the capacity of a subject to determine themselves, i.e. to act without being controlled from outside: a person is spontaneous when their behaviour reflects their inner emotional state in an autonomous way, for example when they laugh or are moved, without feeling obliged by particular social rules to adopt a certain attitude. In scientific language, on the other hand, the concept of spontaneity refers to how a system or physical phenomenon manages its own energy equilibrium in relation to its environment: a system goes through a *spontaneous transformation* when a change takes place without an external force acting upon it and expending energy. This spontaneity is a matter of intimate concern for we humans. Our organism was not built one molecule at a time; the biochemical structures that compose us have formed and continue to reform autonomously, using the matter we absorb from the environment to spontaneously assemble our bodies. Imagine taking any broken machine—a car, for example, or a washing machine—and 'feeding' it with a smoothie made out of its constituent substances: it will never be able to self-heal even the smallest part of its structure. This is because the relations between the parts that make up artificial machines—or at least machines as they are conventionally understood—are not intrinsic. That is to say, they are not *spontaneous*, but are imposed at the time of construction, and exist, we might say, on a separate plane from that of the material of which the machine is composed. But is it possible to imagine an artificial machine capable of growing, reproducing, and regenerating in the same way as our bodies do?

Comparing a machine with a living organism may seem a little harsh. These are two types of systems that seem to obey completely different physical laws: while the organism grows and regenerates itself, the machine as we are accustomed to conceiving of it without human intervention can do nothing but hang around and slowly disintegrate, subject as it is to the effects of the well-known thermodynamic principle that the internal 'disorder' of any isolated system inexorably increases. We might even imagine, following the animistic line of thought, that living organisms possess an immaterial quality that makes them different from the rest of the matter around them, something that enables them to fight upstream against time's merciless flow. For a long time it was thought that the laws of thermodynamics were not sufficient to explain the physics involved in the processes of growth and differentiation exhibited by living organisms; in particular, embryos seemed to respond to principles of differentiation and spontaneous organisation that were totally at odds with the physics of the time. The embryologist Hans Driesch, who lived between the end of the nineteenth century and the beginning of the twentieth century, was one of the last great exponents of the current of thought known as *vitalism*, according to which the organisation of living organisms could not be explained via recourse to the laws of physics alone. Observing the ability of sea urchin embryos to produce an entire adult organism even when they were cut in half, Driesch came to the conclusion that living matter must be animated by an immaterial principle capable of directing its development, which he defined using the Aristotelian term *entelechy*.[1]

In reality, the laws that govern the organisation of living systems and those that lead to the disintegration of artificial systems are exactly the same, and there is no mysterious force that allows living matter to organise itself autonomously. But this realisation forces us to ask ourselves *why* the same physical principles can lead to such apparently opposite results. The difference between ordinary machines and living organisms lies in their different architectures: while machines are generally built following a top-down approach (from top to bottom), i.e. assembled by an external agent from a preformed design of their structure, living systems are formed through a bottom-up process (from bottom to top), i.e. they emerge spontaneously from the continuous and dynamic interaction of their component parts.

The almost miraculous character of the spontaneous transformation from the disorder of inorganic matter to the order of living organisms is well exemplified by myths of creation in which the creator or demiurge builds a perfectly functioning organism from clay or dust. In these myths it is often the divine breath, a spiritual and immaterial force similar to the vitalist entelechy, that makes possible the transition of matter from a completely disorganised configuration to an ordered structure. The matter

---

1.   H. Driesch, *The History and Theory of Vitalism* (London: Macmillan, 1914). *Entelechia* was the term used by Aristotle in opposition to 'power' (*dunamis*), to describe a reality that has reached its highest degree of development.

that makes up the dust and the matter that makes up the organism are exactly the same; the essential difference lies in their organisation. Religious perspectives aside, science has on many occasions asked a similar question, i.e. how it was possible for biological life to first take shape out of a 'broth' of disorganised chemical ingredients. Finding an answer to this question would not only allow us to understand our history—to understand once and for all where we come from—it could also help us to understand whether, and under what conditions, other material organisations, whether natural or artificial, terrestrial or extra-terrestrial, can be considered as living. By shifting our attention from the specific chemicals that make up the body of a living being to the structure that connects them, we may one day be able to build artificial technologies capable of growing and replicating themselves like real organisms. Life itself, if released from its strictest definition, could be a more widespread phenomenon in the universe than we think: material structures very different from us, but capable of spontaneously organising themselves, may well be found swimming in the seas of liquid methane that cover the surface of Titan, or vaulting through the acid clouds of the atmosphere of Venus.

In the Jewish mystical tradition, the possibility of reproducing the process of creation is embodied in the figure of the Golem, an anthropomorphic organism capable of self-assembling from dust at the command of a human being. In the stories of the ancient Kabbalists, the organising force that allows the Golem to maintain its human configuration, keeping it at least apparently alive despite the natural forces that should lead to its disintegration, is a magic word traced on its forehead: the Hebrew word *emet*, 'truth' (or, more literally: what is stable and firm, and therefore true). Once the organism has outlived its usefulness, or threatens to escape the control of its creator, all that is necessary is to erase the word from the Golem's forehead in order to reduce it again to the heap of matter from which it emerged. What I find particularly fascinating in the mythology of the Golem is that the process of its creation and dissolution highlights the connection between matter and information, albeit expressed in magical occult form. Here the animistic idea of the vital breath is replaced by something which, while remaining immaterial, is far more defined and concrete: it is the word, a 'string' of information, that enables matter to organise itself according to a particular structure. Far from being a transcendental principle, however, the Golem's magic word is inscribed in the dust it is made of, and cannot be separated from the creature it animates without completely destroying it. It is interesting to observe that, in modern Hebrew, the word *golem* also means 'robot': and indeed these two beings, one mythological, the other technological, have much in common, being both anthropomorphic artificial bodies capable of obeying the commands of a human being. The difference is that, whereas the Golem is capable of spontaneous self-organisation, the robot, at least in the traditional sense, must be assembled by humans one piece at a time.

Nowadays, the formalisation of the mysterious relationship between material organisation and information is no longer a subject reserved for occultists and sorcerers who spend their days engraving magic words on clay statues. It lies at the heart of the study of what, since the second half of the twentieth century, have been called *complex systems*. In the context of the natural sciences, a complex system, for example a living organism, is a physical system made up of a multiplicity of components in continuous interaction with one another, whose collective behaviour cannot be described simply as the sum of the parts of the individual elements. Frequently, complexity is accompanied by a certain level of self-organisation, which *emerges* from the collective behaviour of the parts. This self-organisation, which steers a complex system away from its thermodynamic 'destiny', contains a certain amount of information, encoded within the *material relations* between the components of the system.

Mathematician Norbert Wiener, one of the most influential thinkers of the twentieth century, founded cybernetics, an interdisciplinary theory oriented toward the study of the self-regulating mechanisms of living systems and artificial machines. In his pamphlet of 1964, *God and Golem, Inc.*, on the relationship between technology and religion, Wiener draws a parallel between the design of new cybernetic machines, potentially capable of learning and self-replicating, and the process via which the Golem is incarnated. As the artificial and 'technological' equivalent of a product of divine creation, the Golem is an ambiguous figure which has often been used to warn of the risks of using knowledge to go beyond the limits of permissible human action. According to Wiener, even contemporary science is limited by such a taboo, a superstitious fear that prevents us from placing living organisms and artificial machines on the same plane. Yet surpassing these limits might allow us to access new levels of technological development. If our fear of the proximity between living organisms and machines could be overcome, it would open up the possibility of building artificial systems with unprecedented capabilities:

> [E]ven in the field of science, it is perilous to run counter to the accepted tables of precedence. On no account is it permissible to mention living beings and machines in the same breath. Living beings are living beings in all their parts; while machines are made of metals and other unorganized substances, with no fine structure relevant to their purposive or quasi-purposive function. [...] If we adhere to all these tabus, we may acquire a great reputation as conservative and sound thinkers, but we shall contribute very little to the further advance of knowledge.[2]

The question Wiener tries to answer by turning to the image of the Golem is far from trivial. Is there a 'missing link' that might relate machines as we have always conceived of them, i.e. as inorganic structures assembled by way of external action, to living organisms capable of spontaneously organising themselves? At the time when Wiener was writing, this was a highly topical and scientifically relevant question. Following different approaches, bringing together physics, mathematics, chemistry, and biology, many scientists from the second half of the twentieth century on tried to formulate theories that could account for the *organisation* of matter both living and non-living, embracing it in its entirety—that is, without reducing it to a simple 'side effect' of mechanical or thermodynamic laws already established for those simple systems traditionally considered by physics.

---

2.   N. Wiener, *God and Golem, Inc.: A Comment on Certain Points Where Cybernetics Impinges on Religion* (Cambridge, MA: MIT Press, 1964), 5.

Thinking Complexity

To date, there is no universally accepted definition of complexity. All definitions, however, agree on the fact that there are two fundamental characteristics of a complex system: a multiplicity of constituent elements, and the presence of non-negligible interactions between them. This definition, although very general, is sufficient to exclude from the picture most of the systems that have traditionally been studied by physics. Newtonian physics, in fact, although able to study systems of bodies interacting with one another, through gravitational or electrostatic force for example, can provide exact predictions only when that interaction is limited to a maximum of two objects. This limit is known as the *three-body problem*, according to which, in general terms, it is not possible to derive an analytical equation of the Newtonian trajectory of three bodies in reciprocal interaction. Statistical thermodynamics, as formalised by physicist Ludwig Boltzmann in the 1870s, made it possible to understand the laws governing heat transfer in macroscopic bodies as the result of the microscopic movements of atoms. By its very nature, therefore, Boltzmann's thermodynamics is concerned with studying systems made up of an enormous number of components. But rather than being followed individually in their own trajectories, here they are treated collectively from a statistical point of view. Boltzmann's thermodynamics paved the way for the study of the processes of the organisation of matter; however, its principles appear to contradict the possibility that a system could spontaneously maintain an organised structure.

Since the 1950s, various disciplinary fields have recognised the importance of making a breakthrough in our scientific understanding of materials. The inability of science to deal with organisms on their own terms, rather than as aggregates of parts that can be treated separately, was, and in part still is, a serious limitation in the

human understanding of the world. In our experience of reality we
continually encounter objects that we intuitively interpret as 'whole'
bodies with a certain autonomy from the external environment in
which they are located. Although the elementary notion that all matter is made up of atoms is universally accepted, we continue to treat
many macroscopic bodies as individuals in their own right, and not
as aggregates of smaller objects. This is generally true for living organisms and inanimate objects alike, and it is especially true for we
who perceive ourselves, in our experience of the world, as integrated wholes. And yet we are quite aware that these bodies that we encounter on a daily basis are composed of smaller parts: organs, cells,
molecules, and atoms that continuously interact with each other to
produce what we perceive as a totality. In other words, our daily
behaviour never takes account of the molecular or atomic structure
of the bodies that surround us: if we stretch out our hand to grasp
a glass, we know that it will be hard, cold, fragile, and transparent
without the need to have a precise knowledge of how the silicon and
oxygen atoms that make it up are arranged. In the same way, if we pet
our cat, we know that she will squint and purr without having to understand the effect of our touch on every cell of her body. This does not
mean that the microscopic mechanisms—atomic, molecular, and cellular—have nothing to do with the macroscopic behaviour of bodies. It
does mean, however, that these are very effective ways of knowing the
reality around us precisely because they treat objects not as a collection
of smaller components but as integrated and coherent sets. And theories of complexity emerged precisely with the aim of giving a scientific
foundation to the intuitive experience of being *an organism in interaction
with other organisms*.

When speaking of complex systems, reference is often made to the
two concepts of *reduction* and *emergence*. According to the philosopher
Edgar Morin, one of the most important authors to promote a *thinking of complexity* that cuts across the sciences and the humanities, 'the
principle of reduction is based on the idea that the knowledge of the
basic elements of the physical and biological world are fundamental,
whereas knowledge of the various changeable groupings into which they

enter is secondary'.[1] This prejudice has led science to try to reduce the complexity of the relations between and within bodies to a set of ever smaller elementary units capable of interacting with each other only locally. The reductive approach is also reflected in the way in which different sciences are commonly placed into relation with one another. One of the most common cliches concerning the natural sciences is that there is a sort of hierarchy of scientific disciplines, in which physics is more fundamental than chemistry, which is in turn more fundamental than biology, and so on down to the human and social sciences. This is an example of everyday reductionism, based on the assumption that the relational structure of complex objects—molecules, materials, cells, organisms, society—does not require any specific knowledge or study in order to be properly understood, but can invariably be reduced to a basic set of simple physical rules.

The concept of *emergence* is often mistakenly invoked in arguing that science cannot provide any explanation for the emergence of higher-order organisational structures in nature, such as the mind or living organisms. According to this interpretation, it is necessary to appeal to a higher principle, external to matter yet capable of determining its organisation. But in reality it is absolutely possible to study complex systems in a rigorous way, and their behaviour can be modelled on the basis of their components, provided that the relational nature that defines them is taken into account. The theoretical biologist Ludwig von Bertalanffy is considered to be one of the founders of complexity thinking; his most important theoretical work formalised the properties of interacting systems of bodies within a *general systems theory* that cut across disciplines, being transferable from chemistry to biology to the study of human social behaviour. Von Bertalanffy provides a very clear explanation of the concepts of reduction and emergence:

The meaning of the somewhat mystical expression 'the whole is greater than the parts' is simply that constitutive characteristics are not explainable from the characteristics of isolated parts.

---

1. E. Morin, *La sfida della complessità* (Florence: Le Lettere, 2017), 33.

The characteristics of the complex, therefore, compared to those of the elements, appear as 'new' or 'emergent'. If, however, we know the total of parts contained in a system and the relations between them, the behavior of the system may be derived from the behavior of the parts.[2]

In general, the most significant emergent property of a complex system is *its very existence as a system*, i.e. its ability to maintain its relational structure intact, without which it would fall back into a disorganised state. From this capacity derives the phenomenon of *self-organisation*, which is the most interesting and technologically most promising feature of complex structures. A system capable of self-organisation is able to increase its level of internal order spontaneously, restoring its structure to within certain limits even when it has been modified by an external force. We have already seen an example of self-organisation in action in the ability of spider silk proteins to construct, transform, and reconstruct their hierarchical structure autonomously. In the 1970s, biologists Humberto Maturana and Francisco Varela generalised the idea of self-organisation in living systems via the concept of *autopoiesis*. According to their definition, 'an autopoietic machine is an homeostatic [...] system which has its own organization [...] as the fundamental variable which it maintains constant',[3] i.e. a machine that has itself as the product of its own function.

The most interesting aspect of the concept of autopoiesis is that, from Maturana and Varela's perspective, the condition of autopoiesis is necessary and sufficient to define life: every autopoietic machine is living, regardless of the specific nature of its constituent components. This generalisation is significant above all because it opens up the possibility of the existence of life forms different from the ones we know of: since the characteristics of the components of an autopoietic machine can in principle be anything, the phenomenon of life is reinterpreted as

2.   L. Von Bertalanffy, *General Systems Theory: Foundations, Development, Applications* (New York: George Braziller, 2015), 55.
3.   H.R. Maturana, F.J. Varela, Autopoiesis and Cognition: The Realization of the Living (Dordrecht and Boston, MA: D. Reidel, 1980), 79.

the result of a relational structure rather than as a specific property of a certain type of matter. In this sense, Maturana and Varela's thought explicitly proposes the possibility of the production of one or more forms of artificial life; like the mythological Golem, it calls into question the rigid boundary between machine and organism.

> There seems to be an intimate fear that the awe with respect to life and the living would disappear if a living system could be not only reproduced, but designed by man. This is nonsense. The beauty of life is not a gift of its inaccessibility to our understanding.[4]

Many of the 'intelligent' behaviours of the materials we have encountered in previous chapters, such as their ability to respond to stimuli, to repair themselves, or to construct a memory of their past, may be understood in the light of theories of complexity. Not only that, but the study and design of new materials makes it possible to construct 'model systems' that provide a direct experimental encounter with the phenomena of emergence and self-organisation that are characteristic of complexity. This is certainly fundamental in trying to understand and reproduce the physico-chemical processes that enabled the emergence of life from disorganised matter; but, more modestly, the complexity paradigm also allows us to understand the more banal and everyday properties of the materials that surround us and which, in turn, themselves *emerged* from a huge number of molecular interactions. Why is a piece of iron hard and a piece of rubber flexible? Why is water wet? Such questions, which may seem rather childish, require us to take a complex view of the objects with which we relate.

By definition, a material is made up of a set of microscopic components—atoms, molecules, particles, fibres—which interact with one another to form a macroscopic object. The word 'complexity' itself brings us back to the example from which our journey began: 'complex' comes from the Latin *cum-plexus*, which literally means 'intertwined' or 'interwoven'. It is therefore evident that the study of materials is strictly

---

4.   Ibid., 83.

related to the study of complex systems: in studying materials, we intuitively understand that they have intrinsic properties not limited to those described by Newtonian physics (i.e. mass, velocity, and position). On the contrary, our perceptual experience communicates to us that there are *hard, soft, elastic, viscous, opaque,* and *transparent* bodies: categories that provide a description of reality that is indispensable and just as effective as that provided by Newtonian trajectories. In other words, while classical physics tells us about a simple universe made up of point masses moving along trajectories, it is only through an understanding of the *relational structure* of the materials around us that we can account for our experience of a dynamic and multiform universe.

Bricks and Atoms

In the course of its development, the thinking of complexity has focused mainly on the study of complex structures that already exist in
nature. The idea of artificially building complex synthetic structures
capable of reproducing the virtues of natural complex structures—i.e.
self-organisation, robustness to change, and response to environmental
stimuli—remained little more than mere speculation until the end of
the last century, at the earliest. It was the development of complexity
and various forms of cyber-theory that finally called into question the
theoretical boundary separating living organisations from machines,
showing on the one hand that a general formalisation of the behaviour
of complex systems was possible and, on the other hand, that such
behaviour could at least in principle be artificially reproduced. Thanks
to the development and popularity of these new fields of study, fertile
ground had also been created for the production of a new interdisciplinary vision of science: the complexity and systems theory approach was
the key to an interpretation that could be applied across the natural
sciences, engineering, and computer science, paving the way for a new
technological renaissance in which rigid barriers between different disciplines would be easily overcome. Between the end of the 1980s and
the beginning of the 1990s, armed with a unified model of machines
and organisms, scientists were prepared to enter the new millennium
as alchemists, capable of transmuting information into matter and inorganic structures into living forms.

Within that historical moment, the main limit to the construction of
artificial structures that could manifest self-organising behaviour similar to that of living matter was above all the *problem of scale*. The phenomena of natural self-organisation characteristic of living organisms
involve structures between the nanometre (nm) and the micron (µm),

i.e. between one billionth and one millionth of a metre; this dimensional range, which may be defined, for simplicity, as the *nanometric scale*, is very close to atomic dimensions: for example, a gold particle with a radius of 2nm contains only two thousand atoms. If that seems like a lot, consider that a 5g gold object contains about 10^{22}—that is, ten trillion, atoms. Self-organisation requires that a great many components act together synergistically: the multiplicity of elements and structural levels contained in a single square centimetre of living tissue is immeasurably larger than any current technology can achieve within that same space. In addition, the interactions that allow a system to self-assemble can exist at any scale, but are more relevant in nanoscale systems because nanometric objects, endowed with a far smaller mass, are less subject to the effects of external forces such as gravity, while the surface available for their mutual interaction is much larger. To understand the counter-intuitive notion that a smaller thing is, at the same time, much larger, imagine a cube with 1cm long edges. Its surface area is initially $6cm^2$, but if we start dividing it into smaller and smaller cubes, its surface area increases enormously. If we divide it into tiny cubes with 1nm-long edges, then with exactly the same volume and the same amount of matter, we will attain an available surface area of $6000m^2$, approximately the size of a football pitch. And most of the chemical and physical interactions in a system take place on the surface—or, better, *at the interface* that separates its components. Increasing the surface also increases the possibility of producing and controlling new forms of self-organisation.

In other words, there is 'plenty of room at the bottom', as Richard Feynman speculated in a now celebrated 1956 lecture where he suggested that the manipulation of matter on the nanometre scale would open up a boundless world of incredible technological opportunities.[1] Indeed, it is clear that, in order to rival the amazing capabilities of natural structures, it is essential to find a way to act on a similarly small scale. Today we use the term nanotechnology to refer to the body of theoretical and practical knowledge that allows us to manipulate matter on

---

1.   R.P. Feynman, 'There's Plenty of Room at the Bottom', *Engineering and Science* 23:5 (1960), 22–36.

the nanoscale. Thinking back to spider silk, for example, it should be immediately obvious that such a light and high-performing material owes its properties to the fact that it is able to contain within a very small volume a huge number of complex interactions. What is perhaps not so intuitive is the reason why constructing nanometre-scale objects is so complicated that it even called for the birth of a new science.

In general, the problem of action on the nanoscale can be divided into two aspects: the first is a problem of manipulation, the second a problem of *replication*. The construction of nanometric structures via the direct manipulation of atoms in the same way as, say, we would build a wall from bricks, faces the obvious issue of manipulating atoms individually and 'forcing' them to stay where we want them to. This is problematic in itself, because almost every tool we might conceive of using to move an atom would itself be made of a number of atoms, and would therefore be too coarse to allow for precise manipulation. On such a small scale, matter is 'sticky': because of the electro-static interactions between atoms, nanometric objects tend to stick to anything, making their manipulation an even more complicated affair. Moreover, atoms never stand still, but move with a certain kinetic energy determined by their temperature; in order to 'capture' them, we would have to work at a very low temperature. Not to mention that the air around us is full of molecules, which we would need to eliminate by creating a vacuum before we started work. Finally, unlike bricks, atoms cannot be 'stuck together' with cement, but interact with one another to form molecules following specific proportions and geometries determined by their electronic structure; it is not possible to build molecules by attaching atoms together in a specific geometry unless the 'rules' of chemistry allow it.

As far as replication is concerned, it is clear that the process of atomic assembly, which involves all the critical issues I have just described, must also be repeated many times before obtaining a macroscopic material, which, as we have seen, would be made up of tens of billions of individual atoms. In principle this is not a completely impossible feat: the individual manipulation of atoms of xenon, a noble gas which is almost completely inert, was first achieved in 1989 by a group of IBM scientists using an extremely sophisticated instrument called a tunnel effect microscope, which operates by inducing a difference in potential between the material sample and a very fine metal tip which is used to move the atoms one by one. This is however an enormously costly approach, is not generalisable to different materials, and would not be practicable on a large scale at all.

Unexpectedly, despite the enormous technological difficulties involved in the precise manipulation of atoms, nanotechnology goes back a long way: the oldest example that has survived to the present day dates back to a time well before the twentieth century. It is a Roman artefact from the 4th century AD known as the Lycurgus Cup: a glass cup that is green, but displays an intense ruby red colour when held up against the light. From the time of its acquisition by the British Museum in 1950, the peculiar chromatic characteristics of this cup had archaeologists scratching their heads, until, at the end of the 1980s, the analysis of a fragment of its glass using a transmission electron microscope revealed the presence of very small gold and silver particles, a few tens of nanometres wide, incorporated into the glass matrix.[2] Precise control of the size and shape of the metal particles in the cup did not, of course, require the ancient craftsmen to move the atoms one by one, but was achieved spontaneously by exploiting the capacity of the metals, when pre-dissolved into the glass, to self-organise into particles with specific morphological characteristics.

It is the presence of these gold nanoparticles that makes the material interact so unusually with light, exhibiting a very complex physical phenomenon known as *surface plasmon resonance*, one of the most interesting behaviours exhibited by matter on the nanoscale. Because the nanoparticles contained in glass are so small, the most energetic electrons in the gold atoms at the surface of the particle manifest a coherent behaviour, forming what physicists call a *plasmon*: the collective oscillation of electrons in a plasma. Like all quantum objects, the plasmon behaves like a wave: when confined to the surface of the nanoparticle, it vibrates only at specific characteristic frequencies.

2.   I. Freestone et al., 'The Lycurgus Cup—A Roman Nanotechnology', *Gold Bulletin* 40:4 (2007), 270.

This particular behaviour has to do with the phenomenon of constructive and destructive interference, characteristic of all waves from electrons to sound waves to the waves that form on the surface of water. If we throw a stone into a pond we observe the formation of concentric waves; if we throw two of them which hit the surface at two different points, we will observe that the waves formed by the two stones interact with one another in a particular way: in some regions of the pond they will enter into so-called constructive interference, adding to and intensifying one another, while in other regions they will cancel one another out completely. The plasmon behaves similarly when interacting with light, which, as we know, is also a wave. When the frequency of visible light corresponds to a response frequency of the plasmon, i.e. when constructive interference takes place, the light of that specific frequency is absorbed while the remaining light is transmitted. Each frequency of visible light corresponds to a certain colour; in the case of the nanoparticles in the Lycurgus Cup, the transmitted light appears red. The transmitted colour will vary depending on the size and shape of the nanoparticles used.

Ancient Roman glassmakers could not have had any idea of the quantum phenomenon of superficial plasmon resonance: they arrived at this result experimentally, perhaps even by mistake. The fact that, despite the technological difficulties involved in the direct manipulation of atoms, the ancients were already able to achieve precise control of matter on the nanoscale, should suggest that there are far more efficient and far smarter ways of approaching nanotechnologies, methods which do not involve the individual assembly of atoms, but instead exploit the capacity of matter to spontaneously organise itself into different structures depending on environmental conditions. The ability of gold to transform itself into a ruby red substance when chemically treated according to certain processes has a quasi-alchemical appeal: perhaps we cannot rule out the possibility that the lairs of the ancient alchemists were, among other things, primitive nanotech laboratories.... Glass produced with similar nanotechnological approaches, which allowed craftsmen to produce a wide range of different colours using very fine metal nanoparticles, already decorated the windows of ancient Gothic cathedrals: they were not anthropomorphic automata like the Golem of the Kabbalists, but were based on a strangely similar principle, which exploited the capacity of apparently inert raw material to organise itself autonomously into fine-grained complex systems. In any case, we are dealing with a practical, essentially experimental knowledge developed in the kilns of craftsmen long before the very existence of atoms was ascertained with any certainty. Once again this is proof that the bottom-up approach, which makes it possible to assemble complex structures by making use of spontaneous organisation from below, allows material to be controlled intelligently.

Because of the enormous posthumous success of his lecture, Feynman is often regarded as the father of nanotechnology—this honour is certainly not bestowed upon some nameless ancient craftsman. In Feynman's time, the word 'nanotechnology' had not yet been invented: he merely describes it as *the problem of manipulating and controlling objects on a small scale*. According to Feynman, the value of this new science lies precisely in the possibility that 'it might tell us much of great interest about the strange phenomena that occur in complex situations', precisely because on the nanometre scale there are an enormous number of simultaneous interactions between many different components of a system. The starting point here, as for the scientists who first formalised the properties of complex systems, was the observation of natural self-organising phenomena: Feynman says that he was 'inspired by the biological phenomena in which chemical forces are used in repetitious fashion to produce all kinds of weird effects (one of which is the author)'.[3] However, in his lecture the physicist seems to focus exclusively on the possibility of direct and individual manipulation of atoms, even considering moving them one by one with 'little hands' directly controlled by an operator through a complicated chain of mechanisms. Once again, at the heart of the problem of complexity we find the question of control: Is it possible to direct the behaviour of a complex system without individually manipulating each of its components?

---

3.   Feynman, 'There's Plenty of Room at the Bottom', 22.

Following in Feynman's footsteps, it was the scientist Eric Drexler who coined the word nanotechnology in a controversial 1987 essay entitled *Engines of Creation*.[4] Drexler's book is a lengthy speculation on the possibility of building nanoscale automata, which he calls *nanobots*, and on the potentially miraculous or catastrophic consequences that this new technology could have upon human civilisation. At the heart of Drexler's theory is the idea of using tiny objects called *molecular assemblers* to solve the problem of manipulating matter on an atomic scale. In Drexler's vision, assemblers would be able to reproduce themselves by means of other nanobots called *replicators*, multiplying exponentially so as to be able to construct any desired macroscopic object in a very short period. The influence of ideas of complexity and self-organisation on Drexler's work are obvious to see: unlike Feynman's 'little hands', Drexler's assemblers are like a complex decentralised system—an *autopoietic* system, we might say—in which control is delegated to a multiplicity of components capable of organising and replicating themselves autonomously. However, the complete lack of any clear indication as to the nature of these nanobots, combined with a fundamental misunderstanding of the chemistry and physics behind complex molecular interactions, made *Engines of Creation* little more than a visionary work of science fiction. Drexler, like Feynman, remained stuck in the idea that atoms would have to be moved one by one in order to build any structure, rejecting any chemical approach to the problem of the nanoscale organisation of matter.

In 1996, American chemist Richard Smalley received the Nobel Prize in Chemistry along with Robert Curl and Harold Kroto for their 1985 discovery of buckminsterfullerene (also known as the *buckyball* or C60), a football-shaped molecule made up of sixty carbon atoms, and the first allotrope[5] of carbon besides diamond and graphite to be isolated in the laboratory. This discovery by Smalley and collaborators is considered a milestone in the history of nanotechnology because it

---

4. K.E. Drexler, *Engines of Creation: The Coming Era of Nanotechnology* (New York: Anchor Books, 1988).

5. In chemistry, allotropes of a certain element are structurally different forms of the same chemical species.

paved the way for a variety of carbon-based nanomaterials, including the well-known graphene, which today are the basis of much of the research in the field. But what is most interesting, from the nanotechnology perspective, is that the complex truncated icosahedron architecture of C60 was not built, as Drexler might have imagined, by assembling the sixty carbon atoms one by one using futuristic molecular tweezers. On the contrary, Smalley's buckyball, provided it is given the appropriate experimental conditions, is capable of assembling itself, so much so that its presence has been detected in the depths of interstellar space, far from any human hand. This ability to self-assemble is certainly not a surprise to chemists, who, throughout history, have learned to exploit it on different scales to build incredibly complex objects. Smalley's discovery was a hint that the key to control over the nanoworld was perhaps closer to traditional chemistry than to a new nano-robotics.

This insight, coupled with frustration at the extraordinary persistence of Drexler's ideas in the public debate over nanotechnology, led Richard Smalley to publish an article in *Scientific American* in 2001 entitled *Of Chemistry, Love and Nanobots*, in which he tore apart Drexler's theories one atom at a time, and in which he presented, at the same time, a particularly romantic view of the chemist's work.[6] 'When a boy and a girl fall in love', Smalley argued, 'it is often said that the chemistry between them is good', and, as in chemistry, this mysterious affinity cannot be forced by simply pushing two individuals close to each other. On the contrary, the bond must be induced with alchemical wisdom, providing the necessary conditions for it to take shape and patiently waiting for the statistical dance of molecular courtship to do the rest. A potentially violent dance, considering that, to synthesise his C60, Smalley bombarded a graphite disk with a laser beam under a supersonic helium flow—but, then again, some love stories are more fraught than others.

Smalley's article was the spark that ignited a furious debate between him and Drexler, and resulted in an exchange of open letters published

6. R.E. Smalley, 'Of Chemistry, Love and Nanobots', *Scientific American* 285:3 (2001), 76.

in the journal *Chemical & Engineering News* in 2003.[7] The debate, starting out in almost deferential tones, ended in one of the most violent arguments in the history of contemporary science. 'You and people around you have scared our children', Smalley concluded, pointing out that 'while our future in the real world will be challenging and there are real risks, there will be no such monster as the self-replicating mechanical nanobot of your dreams'.

The history of nanotechnology has proven Smalley right, at least until now: most science on the nanoscale uses chemical and physicochemical approaches for the preparation and manipulation of new materials. As we can see from the refined nanotechnology of the Lycurgus Cup, there is no need to use 'molecular tweezers' to obtain precise control over the organisation of a complex system; on the contrary, the best way to exploit the self-organising capacity of a material system is to renounce all top-down control and allow its relational structure to emerge in all its complexity. Atoms are not inert bricks to be assembled one by one; matter is endowed with its own inorganic will, which, depending on the approach we choose to adopt, can act either as an inertial force opposing our manipulations, or as a startling intelligence that acts to our advantage. In this sense, perhaps nanotechnology has more to learn from the astute strategies of the alchemists of antiquity than from the control fantasies of twentieth-century scientists.

7.   K.E. Drexler, R.E. Smalley, 'Nanotechnology: Drexler and Smalley Make the Case For and Against "Molecular Assemblers"', *Chemical and Engineering News* 81:48 (2003), 37–42.

Synthesising Complexity

Despite the high hopes placed in nanotechnology in the 1990s, neither Drexler's nanobots nor Feynman's little hands have ever seen the light of day. It is perhaps also for this reason that to a large extent public attention shifted elsewhere, to disciplines such as artificial intelligence which seemed more likely to fulfil the science-fiction promises (and/or threats) of their founders, while nanotechnologies were dismissed as a strange failed experiment in collective imagination. According to Drexler's predictions, nanotechnology could have offered humanity an opportunity to completely revolutionise the way we understand technology, but it would also have opened the way to very serious risks for the planet. One of the apocalyptic perspectives explored by Drexler in his book is the so-called *gray goo* scenario, in which the proliferation of self-replicating nanobots ultimately leads to the destruction of the earth, via the suffocation of all organic life. Apocalyptic scenarios of this kind have undoubtedly also had an impact on science fiction imagery: an example of such a cataclysm can be found in the 2008 film *The Day the Earth Stood Still* (dir. Scott Derrickson), in which a giant anthropomorphic alien robot called GORT (Genetically Organized Robotic Technology), sent to earth to annihilate mankind, can disintegrate itself into a trail of tiny self-replicating insects capable of devouring anything they encounter in their path. While advances in artificial intelligence technologies only seem to reinforce fears that we will soon be confronted with self-conscious computers like those that populate science fiction films, from the cunning HAL 9000 in *2001: A Space Odyssey* to the ruthless Ava in *Ex Machina,* catastrophic scenarios like the grey goo apocalypse do not effectively capture either the real risks or the promises of nanotechnology.

Actual research in nanotechnology, which began in the early years of the new millennium and has flourished over the last two decades, has in fact yielded increasingly promising results, while also helping us to understand the strange and complex organisational phenomena to which Feynman referred in his lecture. The path travelled by those who deal with nanoscience, however, is more similar to that of the ancient glassmakers than to that predicted by the physicists: very often, control on the nanoscale is entrusted to the ability of matter to organise itself as environmental conditions change, rather than being achieved through the precise control of atoms. In nanotechnology we speak of *self-assembly*: the tendency of a set of atoms, molecules, or particles to organise themselves spontaneously into complex structures.

The study of processes of self-assembly has become increasingly central to the development of new materials, because it makes it possible to exploit the relationships between the components of a system in an intelligent way without directing their structure from outside. This is particularly relevant on the nanometre scale, where the problems of manipulation and replication are far more serious. Scientists have learned to distinguish between two different types of self-assembly, known respectively as *static* and *dynamic self-assembly*.[1] Both processes are technologically useful, and both have to do with the formation of complex structures, but they differ in the way in which they manage energy. In general, the development and functioning of life can be interpreted as a set of static and dynamic self-assembly processes that combine to produce all the structures and behaviours of living organisms. A good way to understand the differences between static and dynamic self-assembly phenomena is to use two natural systems as models: the virus and the cell.

Viruses are not living organisms, although very often we describe them using language that seems to imply that they have a kind of vitality. Unlike an inorganic structure such as a crystal, a virus seems to be capable of performing a variety of different functions; in particular, it seems to be endowed with a sort of intentionality that prompts it to manifest a parasitic behaviour toward the organisms it infects in order to continue to replicate itself. In spite of this, however, the virus is far

---

1.   G.M. Whitesides and B. Grzybowski, 'Self-Assembly at All Scales', *Science* 295:5564 (2002), 2418–21.

more similar to an inorganic crystal than to a cell. The protein coating, or *capsid*, which makes up the 'body' of the virus and contains the viral DNA or RNA, forms a static and stable structure; in chemistry it is said to be *at an energy minimum*: it can keep its structure intact without the need to consume any further energy. For this reason, once the host cell has reproduced its components, the virus assembles itself spontaneously into its 'adult' form, and will continue to survive (or rather, continue to lead its *non-life*) so long as this 'suits it' energetically—but it will have no problem in falling apart if environmental conditions render its structure unstable. A lump of sugar behaves in the same way: it might remain in our kitchen cupboard undisturbed and intact for years, until we drop it into a cup of hot coffee. It is precisely for this reason that a virus can remain intact and infectious for a long time on the surface of objects. On the contrary, any living organism, including ourselves, needs to consume energy to 'hold' its structure together: if it stops feeding, it will die and disintegrate.

In 1955, two biochemists, Heinz Fraenkel-Conrat and Robley Williams, conducted a famous experiment on the Tobacco mosaic virus (TMV), demonstrating that it is possible to assemble an infectious virus in a test tube simply by mixing a solution composed of its proteins and RNA.[2] This is one of the very first examples of static self-assembly in the laboratory, and highlights the possibility of exploiting the ability of molecules to interact spontaneously with each other to form complex, functional structures. Like an inorganic crystal, the virus emerges spontaneously from random interactions between its molecular components, which end up settling into the most stable structure available to them, a bit like a golf ball rolling around before settling into a hole in the ground. What makes the virus a particularly interesting system, compared to crystals and golf balls, is the level of complexity in its structure, which is not trivially repetitive but exhibits a particular morphology, icosahedral or helical for example, which characterises it as an 'individual' and determines its function.

2.   H. Fraenkel-Conrat and R.C. Williams, 'Reconstitution of Active Tobacco Mosaic Virus from its Inactive Protein and Nucleic Acid Components', *Proceedings of the National Academy of Sciences of the USA* 41:10 (1955), 690–98.

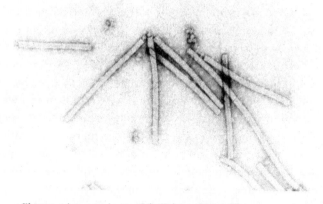

Electron microscope image of the Tobacco Mosaic Virus.

But what makes the virus 'special' compared to any other equilibrium structure is above all the fact that the environmental conditions that determine its assembly and disassembly have been developed by evolution precisely to respond to the specific needs of its reproduction: the virus 'disassembles' inside the cell to reproduce its genome and 'reassembles' spontaneously outside the cell when all of its components have been reproduced, so that it can go on to invade yet another cell.

Clearly, a virus is a natural structure whose components have already been selected and optimised over the course of its evolution. The spontaneous assembly of TMV, from this point of view, may at first seem like the miraculous result of a life force already contained in the matter that makes it up—an *entelechy*—rather than as the effect of a set of complex physico-chemical interactions which, in principle, we are able to understand and control. But the principle of static self-assembly can also be applied in artificial systems. An illustrative example concerns what are known as *targeted drug delivery systems*: this is a class of nanometric objects of a different nature, defined by their ability to carry a medicinal substance and release it within a particular region of the body. Often these systems consist of vesicles of a few tens of nanometres in diameter, made up of molecules capable of self-assembling by exploiting their reciprocal physico-chemical interactions. These molecules encapsulate the drug, and release it by disintegrating in response to a particular stimulus within the target cell. Just as the effectiveness of the virus lies in its ability to assemble and disassemble under precise environmental conditions, these artificial systems use the same principle to, for example, selectively invade and destroy cancer cells.[3]

Nanomedicine, i.e. the use of nanotechnology for therapeutic applications, holds great potential, although it has always struggled to reach widespread diffusion outside of the research stage. Last year, however, the worldwide circulation of m-RNA vaccines against COVID-19 marked a milestone for the application of nanotechnology in everyday life. Although the core of this new generation of vaccines is the m-RNA technology, instructing our cells to produce the SARS-CoV 2 spike protein, m-RNA itself quickly breaks down in the human body, and it requires a carrier to effectively cross the cell membrane. Interestingly, since its early beginnings nanomedicine has been looking at viruses as models of self-organising, natural nanostructures. As a 2020 editorial in *Nature Nanotechnology* observed, 'viruses are naturally occurring nanoparticles, and indeed, the nanotechnology community has long been trying to capitalize on the properties of viruses and mimic their behaviour, for example, for the design of virus-like nanoparticles for targeted drug delivery and gene editing.'[4] Current m-RNA vaccines such as the Moderna and Pfizer-BioNTech vaccines use finely tailored self-assembled lipid nanoparticles to encapsulate, protect and deliver m-RNA inside our cells. In light of this sudden, universal proximity with nanotechnology, the need for a wider and deeper understanding of its guiding principles has become increasingly urgent.

Static self-assembly allows for the development of nanoscale structures with many different features, and is a very smart way to build an object without paying the energy cost of a top-down assembly approach. The vast majority of strategies used in nanotechnology today exploit static self-assembly processes to exert precise control on the nanoscale

3.   Xi Hu et al., 'Biological Stimulus-Driven Assembly/Disassembly of Functional Nanoparticles for Targeted Delivery, Controlled Activation, and Bioelimination', *Advanced Healthcare Materials* 7:20 (2018), 1800359.

4.   'Nanotechnology versus Coronavirus', *Nature Nanotechnology* 15 (2020), 617.

for the construction of complex material systems capable of performing almost any function, from electronics to medicine, from photovoltaic cells to the reduction of environmental pollutants. The variety of structures that can be assembled bottom-up in the laboratory is virtually infinite, from gold particles similar to those discovered in the Lycurgus Cup, to innovative carbon-based materials such as graphene, to organic polymers capable of mimicking the behaviour of biological structures. Provided you can design the right 'building blocks' and identify the correct assembly conditions, the material will be able to organise itself autonomously and perform precise functions in a completely spontaneous way.[5]

Unlike a virus, whose structure, however complex, is the result of stable interactions between its components, a living cell needs to continuously exchange energy with its environment in the form of chemical fuel and heat in order to keep its organisation intact. This is the real meaning of the cliché that every form of life is a system 'battling entropy': that the organisation of matter that constitutes our bodies is not the result of a condition of equilibrium, but that we must constantly maintain a flow of energy through our cells to keep ourselves together. Which in turn means that, in preserving and increasing our internal organisation, we increase disorganisation in the environment around us. It is, however, an oversimplification of the concept of entropy that makes us feel that we are extraordinary thermodynamic exceptions in a universe completely hostile to our existence. Rather than warriors in a desperate cosmic struggle against disintegration, we can think of ourselves, less tragically, as surfers riding the wave of spontaneous chemical processes and exploiting them to our advantage.

This type of far-from-equilibrium organisation, which we can define as *dynamic self-assembly*, is often considered to be a behaviour exclusive to living organisms, but in reality it can be extended to certain specific inorganic systems, one of which in particular is surprisingly simple. In 1900, the French physicist Henri Bénard was the first to observe an interesting behaviour of fluids: a liquid paraffin bath in which

---

5.   C. Huang et al., 'Effect of Structure: A New Insight into Nanoparticle Assemblies from Inanimate to Animate', Science Advances 6:20 (2020), eaba1321.

graphite particles were dispersed, when heated from below, displays a strange hexagonal 'beehive' structure on the surface that disappears spontaneously as soon as the heat is withdrawn. The same behaviour is exhibited by all liquids, including water, and is the result of the spontaneous movement of molecules which, when subjected to a thermal gradient, tend to organise themselves into columns with a hexagonal base, within which they move in a spiral formation from bottom to top. These hexagonal cells of fluid, later christened *Bénard convection cells*, are one of the simplest examples of self-organisation in a far-from-equilibrium system. What makes Bénard convection different from other self-organising phenomena is the fact that, in order for the convection cells to preserve their structure, a heat source must continue to provide a constant flow of energy to the system: when thermal energy is no longer supplied, the cells instantly lose their structure. The chemist Ilya Prigogine defined the structure of Bénard cells, and all other systems whose self-organisation requires a constant flow of energy from the outside, as a *dissipative structure*; it was precisely for his study of the thermodynamics of dissipative structures that he was awarded the 1977 Nobel Prize in Chemistry. In the essay *Order Out of Chaos*, dedicated to the new paradigm of complexity in contemporary science, Prigogine and the philosopher Isabelle Stengers comment on the phenomenon of Bénard cells:

> Classical thermodynamics leads to the concept of 'equilibrium structures' such as crystals. Bénard cells are structures too, but of a quite different nature. That is why we have introduced the notion of 'dissipative structures,' to emphasize the close association, at first paradoxical, in such situations between structure and order on the one side, and dissipation or waste on the other. We have seen [...] that heat transfer was considered a source of waste in classical thermodynamics. In the Bénard cell it becomes a source of order.
>
> The interaction of a system with the outside world, its embedding in nonequilibrium conditions, may become in this way the

starting point for the formation of new dynamic states of matter [...].[6]

The 'new states of matter' to which Prigogine and Stengers refer are above all those that emerge in living organisms, but, as the case of Bénard convection demonstrates, the existence of dissipative structures is not exclusively a feature of living things. When Prigogine first developed the idea of the dissipative structure, only a few examples of material systems capable of self-organising by dissipating energy were known. However, just like the concepts of complex systems and autopoiesis, the concept of dissipative structure is useful primarily because it generalises a property that is typical of living organisms and allows us to identify or design it in other contexts. In this sense, the self-organising behaviour of life is demystified, i.e. it is no longer interpreted as a strange, inaccessible, and almost magical property of a certain kind of matter, but is made available to our technological understanding.

While static self-assembly processes take advantage of matter's ability to self-organise into equilibrium structures, dynamic self-assembly processes such as those that occur in living organisms are based on dissipative structures. In nanotechnology, while static self-assembly has been extensively studied and is used as a strategy to prepare materials for many different applications, the study of dynamic self-assembly is still in its embryonic stages; there are many examples in recent scientific literature of artificial systems capable of maintaining, modifying, and increasing their self-organisation by consuming energy from their environment. The energy used by these dissipative structures may derive from one or more chemical reactions, as is the case with living organisms, which employ a *metabolism*, i.e. a chain of chemical reactions that release energy. But self-organisation can also be powered by other types of energy, e.g. thermal energy, mechanical energy, or a light source. In an article published in the journal *Nature Communications* in 2017, a group of scientists described the collective behaviour of a set of spherical polystyrene nanoparticles dispersed in water and enclosed

---

6.   I. Prigogine and I. Stengers, Order Out of Chaos: Man's New Dialogue With Nature (Toronto: Bantam, 1984), 183.

between two transparent slides.[7] When the dispersion of particles is hit by a laser beam, local temperature differences are created in the liquid, causing the particles to aggregate spontaneously into organised structures, similarly to what happens in the case of Bénard cells. Observed closely with a microscope, these simple aggregates of particles display some surprising behaviours: not only do they spontaneously assemble and disassemble, like any dissipative structure, every time the laser is switched on or off, they also form a great variety of different structures, i.e. with different symmetries, which are able to self-replicate, *self-repair* and even *compete* with one another until the structure that replicates most rapidly prevails over the others present. This is a truly fascinating result: tiny spheres of polystyrene, a common material apparently lacking in any particularly interesting properties, when placed in the right conditions, manifest emergent behaviours usually seen only in living organisms. Another fascinating aspect of dissipative systems highlighted by this experiment is that these structures pose a radical challenge to our idea of determinism—the idea that the same initial conditions will always necessarily lead to the same result. The emergence of several complex organisations from the same initial conditions shows that a minute difference in the initial state of the system—for example, a small disparity in the positions of the particles at the moment of laser irradiation—is enough to allow one structure to prevail over all the others, thus having an unpredictable macroscopic effect on the fate of the system as a whole.

It is important to emphasise that the relevance of these advances does not lie in any immediate technological application: the artificial dissipative structures designed by nanotechnology have not yet had any impact on our everyday lives, and it is likely that it will be some time before such a system is integrated into an item in everyday use. What is significant is that these systems show that it is possible to acquire a theoretical but above all practical understanding of how life develops and maintains its organisation. Nanotechnologies, in this sense, if integrated with the theoretical study of complex systems, can allow us to bridge the gap between our technologies and biological life, responding

7.   Ilday et al., 'Rich Complex Behaviour of Self-assembled Nanoparticles Far From Equilibrium'.

operationally to the fundamental problem of self-organisation. Adopting the perspective of self-assembly is the first small but fundamental step toward better understanding our own nature, designing more efficient machines and—Why not?—accepting that perhaps biological life on Earth is only one particular case of a phenomenon that is far less exceptional than we might think.

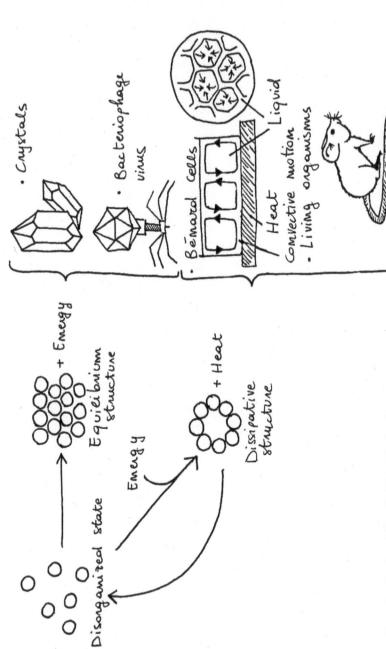

Examples of static and dynamic self-organisation processes.

Back to the Mind

Complexity and the nature of the mind are two problems that overlap at many points. In this chapter I want to highlight how many of the 'intelligent' behaviours exhibited by both living and non-living materials, such as the ability to respond to stimuli and spontaneously assemble their own form, derive from their complex structure. A complex system, by definition, cannot be reduced to its individual components: its structure is essentially relational, so that every element that composes it determines the behaviour of all the other elements that make up the system. This relational structure is a decisive factor in the ability of a material system to respond to environmental stimuli, to preserve its own organisation and also, in one sense or another, to preserve some sort of memory of its own *identity*, which keeps it separate from, but always in communication with, the surrounding environment. In other words, when confronted by the phenomenon of self-assembly in natural and artificial systems, it is legitimate to ask: Who or what is that 'self' that is assembled, and where exactly does it reside? In general, a self-organised system, be it static or dynamic, exhibits what we can call a *circular* response to environmental stimuli. When a complex structure is modified by an external force, if that force is not intense enough to completely disintegrate its identity, the structure responds in a way that allows it to 'return to itself'—that is, to absorb the perturbation that affected it and regain a stable structure, sometimes different from the original one. The subject of the processes of the self-organisation of complex systems is not there *a priori*, i.e. it does not exist 'before' or 'outside' the process; it seems to emerge from the interactions between the constituent parts of the system itself. Perhaps the subject of a self-organising system is completely indistinguishable from the very process of self-organisation that 'holds it together'; in which case all self-organising systems are subjects that perceive their environment, and are all endowed with some form of mind.

Many authors who have contributed to the study of complex systems have sensed the close relationship between thought and complexity, often suggesting that the two concepts were interdependent. Maturana and Varela, inventors of the idea of autopoiesis, define a cognitive system as one 'whose organization defines a domain of interactions in which it can act with relevance to the maintenance of itself'.[1] Emphasis is placed here upon a system's ability to respond in a circular way to stimuli, so as to maintain its structure intact. The mind, therefore, is the result of a *process* rather than an essential property of a specific substance. Complexity thinking once again gives us the resources to generalise a phenomenon such as cognition, which we are used to attributing exclusively to the human, or at most to a small group of mammals, to a variety of systems very different from one another: what counts is the *internal structure* of a system, i.e. the relationships between the parts that make it up, rather than the specific nature of the matter it is constructed from.

But how does an aggregate of polystyrene particles, essentially 'stupid' as they are, 'know' how to organise itself in space according to a precise symmetrical structure, and how to keep that structure intact when damaged? Each particle, taken individually, is no more intelligent than any piece of plastic, but the movement of every single particle has a delocalised effect upon all the others, which in turn influence it again. The result is that, even though none of the particles can know anything about the structure it is helping to form, the whole system thinks and behaves like an organism. As Prigogine and Stengers write, matter, be it organic or inorganic, living or non-living, takes on these 'intelligent' behaviours when it organises itself into *dissipative structures*, i.e. when it exhibits dynamic self-organisation:

[I]n equilibrium matter is 'blind', but in far-from-equilibrium conditions it begins to be able to perceive, to 'take into account,' in

---

1.   Maturana and Varela, *Autopoesis and Cognition*, 13.

its way of functioning, differences in the external world [...] Far from equilibrium fluctuations allow us to use differences in the environment to produce different structures. Once again, we want to emphasise the essential importance of the far-from-equilibrium conditions: 'communication' and 'perception' are the key words for the new behaviour of matter far from equilibrium.[2]

In his 1979 essay *Mind and Nature*, anthropologist Gregory Bateson provides a general definition of mind using the tools provided by systems theory and cybernetics. According to Bateson, 'a mind is an aggregate of interacting parts or components' whose mutual interaction 'is triggered by difference', and whose functioning requires 'collateral energy' and 'circular chains of determination'.[3] This definition can be extended to all complex material systems that are able to respond with a structural change to environmentally-induced perturbations. The concept of *collateral energy* concerns, once again, the dissipative character of these structures: it is incorrect to argue that a dissipative structure 'consumes' energy; rather, it needs to be *traversed* by a flow of energy that is dissipated and immediately returned to the environment in the form of heat; what is exchanged in a relational structure is not energy or matter, but rather information. This also means that a mind, while continually and cyclically returning to determine itself, is necessarily an open system which energy and matter must continually be entering and leaving.

There is a certain cultural paradigm which dictates that the matter that surrounds us is simply made up of small, separate, independent parts that interact with one another by means of simple rigid collisions. At times, this paradigm is so pervasive that it comes to be reflected even in the way we think of individuals within social organisations, as isolated individual elements considered to be acting on their own account alone, without any significant influence passing between them. In light of theories of complexity, we must make room within this atomism for a more dynamic vision of the fabric of reality, one that emphasises the relationship *above* the individual. Too often we think that the only

---

2.   Prigogine and Stengers, *Order Out of Chaos*, 14.

3.   G. Bateson, *Mind and Nature: A Necessary Unity* (New York: Dutton, 1979), 92.

way to design a functioning system is to exercise precise control over each of its components from above. In complex systems, though, it is the fluctuations and deviations, the random and unpredictable movements of matter, that lead to the emergence and evolution of order. It is commonly thought that our mind is a single 'block', an almost magical substance separate from our body. In reality, our ability to behave intelligently depends upon a series of relational and dynamic structures: not only the internal structure of our body and brain, but also the linguistic, cultural, and social fabric in which we are immersed. In this sense, designing complex materials means opening our mind towards new relationships with other minds in order to broaden our cognitive and perceptual potential. Making use of the self-organising processes of complex systems in our technologies would also allow us to revolutionise the way in which we usually exercise control over the materials around us: bottom-up organisation is an extremely efficient way of managing complexity, because it leaves it to matter to build and maintain its own structure. This, of course, requires us to change the way in which we build things; it obliges us to loosen our grip over a matter that we perceive to be stupid and hostile, in order to allow it to best express its boundless creative potential. To technologically exploit the opportunities offered by complex systems, we must accept that the materials we will use in the future may look very different from those we have used in the past. Perhaps in the future we might find ourselves living in fluid houses capable of changing shape along with us like a snail's shell, or in cities capable of disassembling themselves and disappearing only to rebuild themselves spontaneously elsewhere. Although such science-fiction prospects may lie very far in the future, we can be sure that the nanotechnologies of today provide us with all the tools to build and study systems capable of organising, reproducing, and modifying themselves autonomously on different scales, showing us that it is possible to actively imagine and design other forms of life, and new minds.

4

LIVING MONSTERS

I saw how the fine form of man was degraded and wasted; I beheld the corruption of death succeed to the blooming cheek of life; I saw how the worm inherited the wonders of the eye and brain. I paused, examining and analysing all the minutiae of causation, as exemplified in the change from life to death, and death to life, until from the midst of this darkness a sudden light broke in upon me...

Mary Shelley, *Frankenstein*

Nineteenth-century engraving of the scene in Goethe's *Faust* in which an alchemist creates a homunculus. Image: Alamy.

Artificial Lives

Although it may seem like a modern fantasy, the project of artificially creating a living organism already resounds through the experiments of the ancient alchemists. Long before the birth of modern chemistry, the physician and alchemist Paracelsus, in his 1537 volume *De Rerum Natura*, describes a bizarre biochemical procedure for the preparation of what he called a *homunculus*, a synthetic organism with the appearance of a very small human being. Paracelsus's procedure involved the use of certain biological components such as human semen, manure, and blood, which, when subjected to a long process of fermentation, would enable the development of an embryo. While this type of occult operation may seem absurd today, the result of magical superstitions and, in terms of our contemporary view, a fundamental misunderstanding of how life works, such tales of ancient science often contain deeper and more relevant significance than it may seem. Alchemical lore was based on the idea that the inorganic matter encountered by the alchemist in his laboratory was related to the soul of the human being via a complex of mysterious connections and that it was possible, through scientific investigation, to build a bridge between the chaos of inorganic matter and the harmony of the living body, through experimental work in which chemical transformations were understood to also reflect an inner transformation of the alchemist carrying them out. I have always been fascinated by the occult origins of chemistry; I think that, although they are not always evident, contemporary chemistry still bears these ancient influences within it. The idea, so dear to alchemists, of a continuity between the inorganic and the organic world, between dead and living matter, was the basis for the development of modern chemistry and still influences many of our technologies today. But in the ancient alchemical legends I believe we also find symbolic confirmation of the idea that, in chemical synthesis, inorganic matter and human beings exert an influence upon one another, forging a fruitful alliance that makes it possible to produce something completely new. Paracelsus's alchemical procedures, while fascinating, would never have led to any demonstrable result, and yet chemistry, over its long history, has never entirely freed itself from the ambition to discover how to break through from the inanimate realm of dust and crystals to the domain of the living in all its forms.

When Mary Shelley first published *Frankenstein* anonymously in 1818, contemporary science was at a very delicate point. The veil that divided living organisms from inorganic matter was gradually thinning, bringing life and death into dangerously close proximity; increasingly, a certain unexpected *continuity* was being established between the study of biological organisms and other areas of science and technology. We know that in the writing of her novel Mary Shelley was inspired by recent developments in the science of the times. In 1790 Luigi Galvani had observed by chance that the dissected body of a frog was wracked by intense muscle spasms when the nerves of its legs were touched with an electrostatically charged metal scalpel. Galvani sensed that this response had to have something to do with something inherent to biological organisms, a force he called *animal electricity* and which, according to the scientist, was transmitted from the brain to the rest of the organism in order to produce movement.

Today we know that the 'animal electricity' observed by Galvani is the result of the presence of a difference in electrical potential between the inside and outside of our cells, which is called *membrane potential*, and is the result of the selective passage of ions through the cell membrane. Galvani could have had no idea of the nature of the phenomenon he was observing—the very notion of electricity, at the time, was still rather unclear, with electrical phenomena being explained by the existence of an 'electrical fluid' flowing through conductive bodies—but he understood that the body's response to the electrical current must have been related to something inside the body itself. Galvani's experiments also clarified another point: the complex behaviour of life was not simply a problem of energy. It was not enough to provide a dead body with energy in just any form, such as heat or mechanical energy, for it to come alive. In order to obtain responses similar to the behaviour of a living body, something more specific was needed: a *message* expressed in physico-chemical language that the tissues of the organism were able to 'understand'. This specificity is an inescapable characteristic of living matter: unlike the rigid and passive bodies imagined by classical physics, which merely transfer energy by bumping into one another, living organisms seem capable of doing something more: transforming external stimuli into a complex response.

Giovanni Aldini, the nephew of Luigi Galvani, took the study of animal electricity, which he called galvanism, to its extremes, applying his uncle's discovery to the bodies of human beings and exhibiting its miraculous effects in a series of public demonstrations. In his most famous demonstration, which took place in London in 1803, Aldini administered a series of electric shocks to the corpse of a man who had been executed; the man began to come alive, shaken by intense muscular contractions. Aldini's aim was precisely to determine the physical origin of that *vital force* present in all living organisms, and to control it, to the point of crossing the frontier between life and death. The fame of Aldini's demonstration undoubtedly contributed to Mary Shelley's description of Victor Frankenstein's awakening of the Creature, even though there is no trace of lightning or electric shock in her original story: the role played by electricity in the creation of the monster, so central to the subsequent film adaptations of the work, is actually quite marginal in the novel. The legacy of galvanism in Mary Shelley's story is instead mainly linked to the fundamental idea that the miracle of life could become accessible to scientific investigation, and that life itself was not alien to the domain of technological action. In short, the truly frightening aspect of crossing the boundary between living and inert matter was not so much the prospect of animating a corpse as the fear of discovering that we ourselves are but animated corpses, governed by the same forces as inorganic matter, yet somehow alive.

Contrary to what might be expected, although the details of the procedure of creation are never entirely revealed to the reader, in Mary Shelley's imagination Frankenstein's creation of the monster is more like a process of chemical synthesis than the result of a series of intense electrical shocks. The young Victor Frankenstein, with no access to the most up-to-date theories of natural philosophy, begins his autodidactic 'scientific' training following in the footsteps of Renaissance magicians and alchemists such as Paracelsus and Cornelius Agrippa. Upon arriving at university, the young Frankenstein, disappointed by the rigours of a modern science that spurns his dreams of greatness, is driven by a fatal attraction to chemistry, the only scientific discipline that seems capable of fulfilling the miraculous promises of the ancient occult sciences. As his chemistry professor tells him:

> The ancient teachers of this science [...] promised impossibilities and performed nothing. The modern masters promise very little; they know that metals cannot be transmuted and that the elixir of life is a chimera but these philosophers, whose hands seem only made to dabble in dirt, and their eyes to pore over the microscope or crucible, have indeed performed miracles. They penetrate into the recesses of nature and show how she works in her hiding-places. They ascend into the heavens; they have discovered how the blood circulates, and the nature of the air we breathe. They have acquired new and almost unlimited powers; they can command the thunders of heaven, mimic the earthquake, and even mock the invisible world with its own shadows.[1]

These fatal words, which will lead the young Victor down the road to ruin, seem to suggest that chemistry, despite the apparent humility of its methods, is the only science capable of revealing the secrets of life and death to the devoted scientist. And indeed, throughout its history, chemistry has often found itself wandering the edgelands of life, trying to understand the characteristics that a chemical system must possess in order to be considered living. As we have seen, the study of complexity reveals that there is no rigid boundary between living and non-living matter: the difference between these two states of matter

---

1.   M. Shelley, *Frankenstein* (Ware: Wordsworth Classics, 1992), 38.

does not lie in the specific nature of their components, but has to do with different ways in which the same chemical components may be related to one another. What is more, the construction of chemical and nanotechnological systems capable of self-organisation shows that, between the almost completely passive behaviour of a stone (the chemical structure of which, however, still has a certain degree of complexity) and that of an intelligent animal such as a human, there is a very dense spectrum of different material structures, each capable of interacting in a specific way with the reality around it. Life is first and foremost a problem of organisation, and since the dawn of its history, chemistry has always been concerned with understanding how material relates to itself, organising itself spontaneously into the incredible variety of natural and artificial structures that exist. Even had he lived to the present day, the young Victor Frankenstein would probably have found that the study of chemistry answered best to his ambitions; perhaps, however, his monster would not have been a giant two-and-a-half-metre-tall humanoid, but a microscopic organism a few tens of nanometres across, with a far more alien, if no less threatening, appearance.

Shortly after the publication of *Frankenstein* in 1828, one of the most famous experiments in the history of chemistry was conducted. Studying the synthesis of ammonium cyanate from cyanic acid and ammonia, Friedrich Wöhler observed the unexpected precipitation of a white crystalline substance that had a different appearance from the inorganic salt he had expected to obtain. Wöhler identified this substance as an organic compound known as urea, a molecule naturally present in animal urine; he also noted that the substance he was trying to prepare and the substance he had obtained had the same elementary composition, i.e. they were both made up of the same atoms, but arranged in a different pattern. This phenomenon, known in chemistry as *isomerism*, highlights once again the essential importance of *structure* in determining the behaviour of a chemical substance: here too, what separates a 'mineral' substance from a 'biological' one is not the substance itself, but its organisation and structure. Today Wöhler's experiment is remembered as one of the crucial moments in the development of contemporary chemistry, and is often seen as coinciding with the birth of modern organic chemistry as such: it was one of the first occasions upon which a chemical compound produced spontaneously by living organisms was artificially synthesised in a chemical laboratory from inorganic reactants.[2] Prior to Wöhler's synthesis, the adjective 'organic' was used to connote only substances derived from plants or animals; it is thanks to his experiment that today, in chemistry, the word denotes a specific class of carbon-based chemical compounds, and is now devoid of any necessary association with living organisms.

From Wöhler's experiment onward, the ability of chemistry to produce substances that are hybrid, i.e. neither entirely inorganic nor entirely living, called into question the rigid separation between life and death but also, and perhaps more significantly, between life and technology. And this unexpected continuity expressed in the synthetic approach of chemistry also emerges in the way in which the development of the chemical sciences has influenced the language we use to talk about matter. The customary terminology we have used to distinguish the living world from the non-living has always been problematic. Non-living materials have been defined as 'inorganic', 'inert', 'inanimate', or 'disorganised', but each of these terms has ultimately proved inadequate as a way to specify the precise boundaries where life begins: we have synthesised organic materials from inorganic substances and discovered materials that, although not alive, are capable of moving, growing, and remembering. Chemical matter, whether organic or inorganic, is active and dynamic, capable of forming complex organisations on different scales, evolving, and spontaneously modifying its structure in response to the environment. For this reason, rather than considering life as an extraordinary phenomenon, miraculous and extraneous to all other behaviours of matter, chemistry provides us with the tools to think of life as one of the many different types of dynamic organisation that matter can assume. In light of these discoveries, it is still possible to provide a definition of life? How can we draw new boundaries for a category that is so elusive, yet so fundamental for us?

---

2.   See P.J. Ramberg, 'The Death of Vitalism and the Birth of Organic Chemistry: Wöhler's Urea Synthesis and the Disciplinary Identity of Organic Chemistry', *AMBIX* 47:3 (2000), 170–95.

The synthesis of urea is undoubtedly a far more modest result than the one that an ambitious Victor Frankenstein envisioned: the experiment did not lead to the synthesis of an entire organism, it merely demonstrated that certain chemicals present in living organisms can be obtained from inorganic components. However, Wöhler's result can also be regarded as the first step in a long scientific journey that has brought chemistry closer and closer to an operational understanding of living matter.

In 1862 Louis Pasteur definitively demonstrated the impossibility of the spontaneous generation of living organisms from decaying matter, finally putting paid to a deep-rooted belief that had survived in the history of scientific thought since the time of Aristotle. A new principle was therefore emerging: life could only be produced from other life, it was not possible for it to emerge from non-living matter. Two opposing tendencies converged: on the one hand, organic chemistry was increasingly effective in demonstrating the ability of inorganic matter to transform, under the appropriate conditions, into the chemical ingredients of life; on the other hand, biology was now convinced that it was impossible to generate life under any circumstances except from an already-formed living being. But then how was it possible for life to have emerged on Earth in the first place? Pasteur's dogma, while useful in disproving an unfounded superstition about the spontaneous birth of certain organisms, raised a new barrier between inorganic matter and living matter that was destined to be gradually eroded by new scientific theories of the origin of life. The idea that life originated from non-biological chemical components is known as *abiogenesis*, and was advocated by the Russian biologist Aleksandr Ivanovič Oparin, who, in his 1924 book *The Origin of Life*, was the first to develop the theory of chemical evolution, i.e. the idea that life emerged from increasingly complex organic molecules. According to Oparin, this approach makes it possible to overcome both the notion that life is a unique and unreproducible phenomenon *and* the ancient theory of spontaneous generation, rooted in vitalism and equally unviable as an explanation for the origin of life on Earth.

Following in the footsteps of Oparin's theories, the American chemist Stanley Miller, under the guidance of his professor Harold Urey, developed a crucial experiment in 1953, designed with the aim of demonstrating that abiogenesis was indeed possible, and that the primordial Earth could have furnished appropriate conditions in which life's constituent molecules could have been spontaneously produced from a set of inorganic ingredients. Miller developed a simple experimental apparatus consisting of a flask of boiling water in contact with a special gas mixture consisting of ammonia, methane and hydrogen, which, according to the theories of Oparin and Urey, approximated the primordial atmosphere present on Earth at the time when life first emerged. Electrical discharges similar to lightning strikes were then produced within the system by means of two metal electrodes.[3] After a week, Miller purified the aqueous solution he obtained and subjected it to a crude chemical analysis—and confirmed, to his own surprise, that it contained several amino acids, the basic chemical ingredients of living organisms. Beyond the incredible scientific value of his discovery, there is something romantic in Miller's experiment, an echo of the nineteenth-century myth of dead matter animated, in the primordial night of the world, by the power of thunder and lightning.

3.   S.L. Miller, 'A Production of Amino Acids Under Possible Primitive Earth Conditions', *Science* 117 (1953), 528–29.

Inorganic Organisms

Miller and Urey's discovery completed the work begun by Wöhler in the nineteenth century, eliminating once and for all any doubt that the biological molecules that make up living organisms could have been generated spontaneously under particular environmental conditions. However, a mixture of molecules, however complex, is not in itself sufficient to produce life. There is something more to living organisms, in comparison to the primordial broth from which they emerged: they are endowed with a particular *form* that emerges and changes over *time*. Life is an organisation of matter in space and time, the evolution of which is designated by the term *morphogenesis*, the process by which a form grows and changes. While the synthesis of organic molecules in the laboratory suggests that the chemical ingredients of life can be produced from inorganic substances, this fact gives us no indication as to how these ingredients interact with each other in order to create the enormous variety of different dynamic structures that populate the living world. In the process of artificially reproducing life, form and substance present two complementary and indissoluble aspects of the same problem. The first attempts to develop artificial life forms in the laboratory date back to the second half of the nineteenth century, and were focussed on the artificial replication of the processes of morphogenesis observed in living organisms. At that time, all known chemical processes took place in homogeneous mixtures of reagents in solution, and rapidly evolved towards a state of equilibrium which, once reached, meant the system was completely stable and ceased to change over time. An everyday example of this type of chemical reaction would be the dissolving of an effervescent aspirin tablet in a glass of water: as the tablet dissolves, acetylsalicylic acid (the active ingredient of aspirin) reacts with sodium bicarbonate added as an excipient, producing bubbles of carbon dioxide. Once the tablet is dissolved, the glass contains a homogeneous and completely stable solution in which no new bubbles are formed and no further visible development is observed. Chemical processes of this type stand in clear contrast to what we see in living systems, which grow continuously throughout their life, retaining their complex structure made up of several distinct components, each separate from the others. The challenge of chemistry, then, was not only to produce the same molecules as life, but also and above all, the same forms: structures capable of developing in space and time in a manner similar to living matter. The only inorganic structures known at the time that were able to 'grow', i.e. crystals, seemed to evolve in a very different way from living organisms: in crystals the growth impulse comes not from within but from without, through the subsequent deposition of layers of matter on their surface. What is more, the growth of a crystal is almost entirely predictable, being determined entirely by its chemical nature; this makes it largely 'insensitive' to its surroundings. In contrast, the development of a living organism such as a plant or algae is not entirely predictable and is influenced decisively by the environmental conditions in which it grows.

The Polish chemist Moritz Traube was the first to prepare in the laboratory a chemical system capable of mimicking the development of a living organism. In a series of experiments carried out starting in 1867, Traube created what he called 'artificial cells' or 'inorganic cells': dynamic tubular chemical structures with a surprisingly 'living' appearance, reminiscent of primitive underwater organisms or bizarre alien plants. Traube's 'cells' emerged spontaneously from an aqueous solution containing certain chemicals, and displayed certain capacities for growth and self-repair similar to those observed in plants.[1] These curious artificial organisms depend upon a relatively simple chemical mechanism: a 'seed' of a water-soluble solid substance A is dissolved in an aqueous solution containing the chemical B. As it dissolves, substance A reacts with substance B in the solution, forming a semi-permeable solid membrane around the 'seed' whose properties are very

1.   D. Liu, 'The Artificial Cell, the Semipermeable Membrane, and the Life that Never Was, 1864–1901', *Historical Studies in the Natural Sciences* 49:5 (2019), 504–55.

similar to those of cell membranes in living organisms. The membranes thus formed can be penetrated by water but not by other substances; because of the difference in concentration of A between the outside and the inside of the membrane, water is pulled into the inside of the membrane.

The membrane's swelling and, ultimately, its breaching, is caused by osmosis.[2] Once the membrane is broken, the sudden contact between A and B sets in motion its immediate regeneration—resulting in the progressive 'growth' of the cell until the seed is completely dissolved. The result is a solid, three-dimensional membrane, several centimetres high, which expands and gradually grows in water, a bit like a strange marine organism.

While the first artificial cells were made from a small mass of gelatine immersed in a solution of tannic acid, i.e. using two organic substances of natural origin, Traube later reproduced the same results using copper sulphate and potassium ferrocyanide, two completely inorganic salts. This showed that the formation of Traube cells did not strictly depend on the chemical nature of the cells used: inorganic matter could give rise to spontaneous morphogenetic processes in the same way as biological substances. In a letter addressed to Charles Darwin, Traube compared his results in the creation of artificial cells to the theory of evolution by natural selection:

> Your successful endeavour to free the complexity of organic nature
> from the miracle of many particular creations and to trace it back
> to natural causes is clearly closely related to that school of natural

2.   Osmosis is a phenomenon that manifests itself in a solution in the presence of a semi-permeable membrane, i.e. a barrier that can only be crossed by water molecules but not by other substances, for example the ions of a dissolved salt. Spontaneously, a chemical system tends to balance out the concentration of all substances in a solution at every point. In the absence of a membrane, this is done simply by the diffusion of molecules or ions. If, however, there is a membrane that prevents chemical substance from diffusing, the concentration will be balanced by the passage of water through the membrane. Thus, if the concentration of a certain substance is higher inside a membrane than outside, water will spontaneously migrate through it to lower the concentration, causing it to swell and potentially rupture.

science that endeavours to demonstrate that processes considered to be specific to life are simply physico-chemical processes. [...] In this sense one could conclude from my investigations that the organisms which first appeared with cells surrounded by a membrane did not receive this ability to form cells as a new power, but rather borrowed it from inorganic nature.[3]

However, although Traube may have appreciated the significance of his experimental enterprises, the scientific community of his day welcomed them with a fair amount of scepticism. To them, Traube's experiments exploring the capacity of inorganic matter to produce organised and dynamic structures seemed like mere attempts to simulate living forms without actually understanding their authentic mechanisms. But in the early twentieth century, Traube's legacy was embraced by the French biologist Stéphane Leduc, who, on the basis of the Traube cell formation mechanism, conducted a series of experiments in which, using various metallic salts immersed in a solution of sodium silicate, he succeeded in reproducing an incredible variety of inorganic structures that looked surprisingly similar to plants, algae, and fungi. Leduc intended these marvellous 'chemical gardens' to rehabilitate the now ostracised theory of spontaneous generation, reconstructing the lost continuity between living matter and inorganic matter and showing that both could form and evolve on the basis of the same principles of self-organisation.[4] Contrary to what this might superficially suggest, however, Leduc was no vitalist: his belief in the ability of inorganic matter to organise itself was based on solid physico-chemical principles and on experimental observation. For his ideas to be accepted, Leduc believed, a paradigm shift in the biological sciences was necessary: unconvinced by the purely descriptive approach of his contemporaries, Leduc advocated the need for a *synthetic biology* which, by reproducing

---

3.   Traube to Charles Darwin, Breslau, 2 March 1875 (Letter no. 9878), in F. Burckhardt et al. (eds.), *The Correspondence of Charles Darwin* (Cambridge: Cambridge University Press, 30 vols., 1985–), vol. 23, 93–95.

4.   R. Clément, 'Stéphane Leduc and the Vital Exception in the Life Sciences', 2015, <https://arxiv.org/pdf/1512.03660.pdfarXiv:1512.03660>.

One of Leduc's 'Chemical Gardens', from *The Mechanism of Life*, 1911.

the structures of living organisms in the laboratory, would obtain a full understanding of the physicochemical principles underlying their formation. Leduc expresses this idea as follows in his 1911 book *Biologie synthetique:*

> Until now, biology has relied solely on observation and analysis. This sole recourse to observation and analysis in the absence of any synthetic methodology, is one of the factors holding back biology. The analytical method in biology is paralysed, sterilised, by the indissoluble unity of phenomena: if, in a living system, one tries to isolate one phenomenon from the others, then the phenomenon disappears, the animal dies. Not only is the synthetic method applicable to biology as it is to other sciences, but it seems to be the most fertile, the most capable of revealing the physical mechanisms of life, the study of which has not yet begun.[5]

At first glance, Leduc's proposed approach to biology may seem rather crude: How is it possible to really understand something about living

---

5.   S. Leduc, *La biologie synthétique* (Paris: A. Poinat, 1912), 11.

organisms by building inorganic systems that mimic their forms? Isn't there a risk here of confusing form and substance? Yet, in light of theories of complexity and self-organisation, Leduc's thinking seems quite prescient. Precisely because he recognised biological systems as complex, i.e. as irreducible to an unstructured set of simple ingredients, Leduc understood that a radically new approach was necessary in order to study them. If the analytical study of systems involves their separation, either physical or conceptual, into many small independent parts, in the synthetic approach the different components of a system are purposely placed in interaction with one another, allowing the relational fabric that unites them to emerge in all its complexity. In this way the focus is shifted from the specific chemical components of life—amino acids, nucleic acids, phospholipidic membranes...—to the *relationships* that place them in communication with each other and the *processes* that enable their self-organisation. In this sense, even if Traube cells and Leduc's chemical gardens have nothing truly biological about them, they can still contribute to our understanding of life thanks to their ability to reproduce processes and relationships. 'The ordinary physical forces,' Leduc wrote in 1911, 'have in fact a power of organization infinitely greater than has been hitherto supposed by the boldest imagination.'[6] Advances in materials science and nanotechnology have amply vindicated Leduc on this point.

The physicochemical processes that determine the formation of the artificial cells designed by Traube and Leduc are really surprisingly similar, despite the differences, to those that regulate biological structures. In the previous chapter I wrote about different systems capable of self-organisation and we saw how, setting out from the comparison between the virus and the cell, a fundamental difference between static and dynamic self-organisation processes emerges. While static self-organisation, which characterises non-living structures such as viruses and crystals, manifests itself in chemical systems in equilibrium, dynamic self-organisation occurs far from equilibrium and is always fuelled by the dissipation of some form of energy. The inorganic

6.   S. Leduc, *The Mechanism of Life* (London: Heinemann, 1911), 167.

structures that form in Leduc's chemical gardens are driven by irre-versible diffusion processes that occur far from equilibrium, and their growth can undoubtedly be classified as a self-organising dynamic pro-cess similar to those that occur in living organisms. When we speak of equilibrium in chemistry, we do not mean a situation of calm in which nothing happens and time stands still: in conditions of equilibrium physicochemical processes continue, but take place in a completely re-versible way, in a sort of eternal present without future and without memory. Out-of-equilibrium systems such as Leduc's gardens inhabit a temporality more similar to our own: not only do they develop in a specific temporal direction, they are also sensitive to small oscillations of the environment around them, which leave indelible traces in their structure and contribute to the construction of their history.

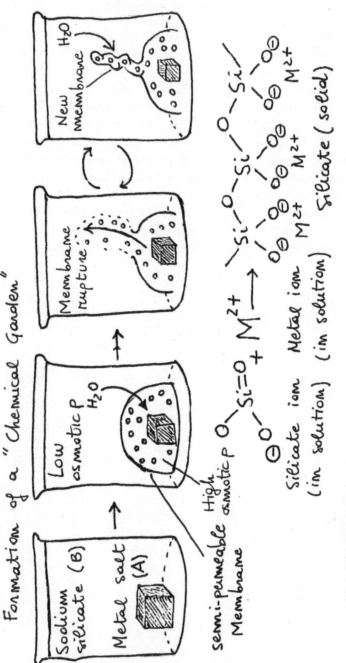

Schema for the formation of Leduc's chemical gardens.

Other Forms of Lyfe

In a recent article published in the journal *Life*, astrobiologists Stuart Bartlett and Michael Wong proposed a new generalised definition of the concept of life.[1] In their essay, the scientists define a new category of chemical processes which they call *lyfe*, broad enough to include a greater variety of 'life as we don't know it'. According to the authors, although terrestrial life is a physico-chemical phenomenon that manifests itself in the presence of liquid water, is based on organic carbon-based molecules, and maintains its internal order through the dissipation of energy, there may be many other forms of life, or rather lyfe, with very different characteristics. Their definition of lyfe is articulated around four fundamental characteristics: *dissipation, autocatalysis, homeostasis, and learning.*

Dissipation, which we have already discussed, is the characteristic of those so-called dissipative systems in which order is maintained through continuous consumption of energy: these are 'open' thermodynamic systems which never reach a state of equilibrium, but are continuously traversed by a flow of energy. Our cells are an example of a dissipative system but, as we have seen, dissipation can also occur in completely inorganic systems, as in the case of Bénard convection cells, for example. *Autocatalysis* is an essential characteristic of molecules that are able to reproduce themselves. A *catalyst* is a chemical substance which, by participating in a reaction, facilitates its development, but is not consumed like a reagent. An *autocatalyst*, then, catalyses its own synthetic reaction, i.e. it acts as a model for its own replication. In some cases, an autocatalyst, in the process of making a copy of itself, transfers information about its structure to the 'daughter' molecule, in a

---

1.   S. Bartlett and M.L. Wong, 'Defining Lyfe in the Universe: From Three Privileged Functions to Four Pillars', *Life* 10:42 (2020).

sort of mechanism of molecular inheritance. And *homeostasis* is a typical feature of complex systems that allows them to keep their internal conditions stable even under environmental disturbances. Finally, *learning* allows the 'lyving' system to acquire information about its environment and to keep track of it within its own structure, as is the case for terrestrial life in the case of Darwinian evolution, in which information about the environment is 'stored' in the genetic code. In the words of Bartlett and Wong:

> By basing the criteria for lyfe on generic processes—rather than specific components that perform specific tasks—we open our minds to the exploration of all systems that display these emergent properties, freeing ourselves from the restrictions of precise chemical recipes whose prescriptions contain assumptions that may limit our explorations of the emergence of life-like behavior in the universe.[2]

Above we have retraced some of the most important moments in the history of modern science, in which the boundary between living and non-living matter has been radically pushed aside, proving inadequate. This process of the gradual deconstruction of the category of living matter has also been accompanied, as we have seen, by a necessary change in terminology, which in turn has had to become hybrid and ambiguous like the matter it is attempting to define: words such as *organic* and *inorganic* have lost their original meaning, based on a metaphysical distinction between life and non-life, and have instead become operational definitions, useful for *building relationships,* rather than barriers, between the countless dynamic states that matter can assume. In light of this scientific and linguistic evolution, a generalisation of the very definition of life—or lyfe—is the natural continuation of a cultural process that allows us to place ourselves in ever-greater continuity with the rest of the matter around us, abandoning the perspective that sees life as an exception, to instead understand it as one of the many dynamic

2.   Ibid., 13.

processes in which chemical matter can participate. It may seem that the idea of founding a 'synthetic biology', rather than a respectable project of scientific investigation, is the unrealisable dream of a modern Dr. Frankenstein, but in reality, in recent years efforts have been made in the direction of building materials that exhibit characteristics similar to those observed in living organisms. The objective of this enterprise is twofold: on the one hand, drawing inspiration from the world of the living in the preparation of artificial materials would allow us to build increasingly intelligent technologies, i.e. technologies more capable of adapting, replicating, growing, and learning new behaviours. On the other hand, following in the footsteps of visionary scientists such as Stéphane Leduc, building artificial systems increasingly similar to biological systems can help us investigate the processes that enabled life on Earth to emerge in the first place, and help us understand the essential characteristics that a material system must have in order to be considered living. In its interrogation of the constitutive characteristics of biological systems, contemporary science is moving more and more in the direction of a *generalisation of the concept of life*. Rather than a specific phenomenon that occurred only once, here on our planet, life is being interpreted as a broader class of self-organising processes that may occur in many different environments, and on the basis of chemical components completely different from those that make up our own bodies. The answer to Enrico Fermi's famous paradox, which asks why we have not yet found traces of life outside our planet, may lie precisely in the fact that we have not yet learned to recognise life forms that are truly different from our own.

Various definitions of life have been put forward by the scientific community. According to the definition provided by NASA, which serves as a reference for identifying the presence of traces of extraterrestrial life, life is 'a self-sustaining chemical system capable of Darwinian evolution'.[3] This definition works fine from a descriptive point of view, but it is not particularly useful from an operational point of view, i.e. it does not provide many indications as to how the properties of

---

3.   NASA, 'Life Detection'. *Astrobiology at NASA*, <http://astrobiology.nasa.gov/research/life-detection/>.

the self-sustaining and evolutionary that characterise life might actually emerge from inorganic matter. From an operational point of view, many definitions of life focus on the fact that a living system can be identified by the co-occurrence of three fundamental functions: (1) the presence of a metabolism, i.e. the ability to consume and use energy, (2) segregation, i.e. the separation of the system from its environment, and (3) the capacity for self-replication.

None of these three functions considered independently is sufficient in itself to define life; they must all be integrated into one and the same system in order for it to be considered living. A certain level of generality immediately emerges from these definitions: biological life on earth may be the only known chemical system currently capable of exhibiting all three characteristics, but this does not necessarily mean it is the only one. Moreover, chemistry and materials science's growing interest in the complexity approach and the rapid development of new techniques in nanotechnology offer science a growing number of tools with which to reproduce these three functions in entirely synthetic organisms. While a perfect integration of these three functions has not yet been achieved in the laboratory, there are many examples in recent scientific literature to show that their realisation in artificial systems is far from impossible. In particular, in a recent study the two properties of *segregation* and *metabolism* were incorporated in a test-tube chemical system in which small, self-organised vesicles were spontaneously generated from a solution of organic molecules, and maintained their structure intact by consuming a chemical 'fuel'.

The study reports on the preparation of an artificial system consisting of tiny nanometre-scale vesicles made up of particular molecules called *surfactants*. Surfactants are organic molecules consisting of a 'head' with an electrostatic charge and a 'tail' made up of a long carbon chain. Because of the electrostatic interactions between the molecules, surfactants have the ability to self-organise spontaneously in water into spherical vesicles by means of a self-organising process similar to that which characterises the membranes of our cells. In this study, the self-organisation of vesicles was made possible by the presence of a particular organic molecule, ATP, which acts as the 'fuel' in living organisms and which, similarly, allows the vesicles to maintain their structure intact. ATP is capable of acting on the heads of surfactant molecules, forcing them to stay aligned and thus allowing the formation of the vesicle. When ATP is degraded by an enzyme added specifically to the solution, the vesicles lose their organisation and must consume a new ATP molecule in order to 'survive'.[4]

This is one example of how two of life's three essential functions, metabolism and segregation, can be achieved artificially and integrated into a single, fully synthetic chemical system. The vesicles produced in this experiment are in fact tiny 'organisms' separated from their environment by a semi-permeable membrane, reminiscent of those primitive synthetic cells prepared in the nineteenth century by Moritz Traube, although chemically far more complex and prepared under far more stringently controlled conditions. Once again, here we are faced with a chemical system capable of self-organisation, i.e. one that increases its level of 'order' spontaneously, starting from a 'broth' of disorderly molecular components and giving rise to structures and complex behaviours. And in this case as in that of biological life, order is maintained thanks to a *dissipative structure*, i.e. through the consumption of chemical energy.

---

4.   S. Maiti et al., 'Dissipative Self-Assembly of Vesicular Nanoreactors', *Nature Chemistry* 8 (2016), 725–31: 725.

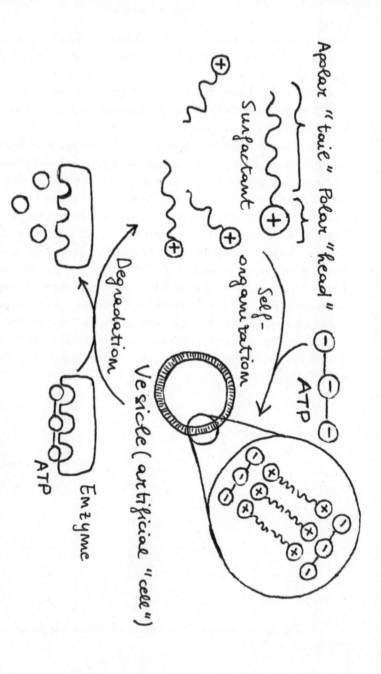

Formation of a self-assembling, energy-dissipating artificial system as described in S. Maiti et al., 'Dissipative Self-Assembly of Vesicular Nanoreactors'.

Life and Information

The ability to self-replicate is often regarded as the single most definitive feature of living matter. For a chemical system to be able to replicate itself, as we have already seen, it must be able to perform a function that in chemistry is known as *autocatalysis*: this is the ability of a certain molecule to reproduce itself starting from its own components, using itself as a model for a copy of itself. In carrying out this process of autocatalysis there must be some kind of transfer of information from model molecule to daughter molecule, which is achieved through the physico-chemical interactions that guide the copying process. This process is typical of biological molecules, but it is not exclusively limited to them: some artificial molecules capable of manifesting self-replicating behaviour have been synthesised, and the search for new chemical systems capable of self-replication is constantly expanding.

One much-cited case is described in an article published in the journal *Science* in 2010, in which a group of scientists constructed two artificial systems capable of self-organisation and self-replication.[1] These particular chemical systems are capable of autonomously self-synthesising from the molecular 'bricks' that make them up, forming fibrous structures of nanometric dimensions visible under an electron microscope. But in addition to the ability to reproduce themselves spontaneously, these self-replicating structures also exhibit another fascinating characteristic: since both are made up of the same building blocks, if confined within the same environment—in this case, the same test tube— the two structures will compete in a similar way to two animal species competing for the same food source. Researchers have shown that, in this in-vitro evolutionary contest, it is not always the same chemical

---

1.   J.M.A. Carnall et al., 'Mechanosensitive Self-Replication Driven by Self-Organization', *Science*, 327:5972 (2010), 1502–6.

substance that succeeds in taking over the test tube: depending on how the solution in the test tube is agitated, one or the other structure will predominate, highlighting how 'selective pressures' from the external environment can influence the evolution of the system in a similar way to what happens in the evolution of living organisms.

Self-replication, on the other hand, among all the functions of living organisms, has long been the most elusive and most difficult to reproduce technologically. It is also often regarded as the most important characteristic of life, because it lies at the basis of the mechanism of evolution: only through self-replication is it possible to transfer genetic information from one generation to another, while at the same time ensuring that different individuals compete with one another in order to establish dominance over others. This evolutionary potential, as demonstrated by the experiment described above, can no longer be considered as the exclusive property of living organisms, but can also be found in relatively simple molecular systems; a fact that has led many scientists to hypothesise that life originated from such a mechanism of molecular self-replication. However, the central role reserved for self-replication processes in the definition of life also has another side to it: overlooking the numerous cooperative processes that are essential to living systems, processes in which collections of molecules and individuals organise themselves to perform complex functions, life is often reduced to an almost disembodied mechanism for the transmission of information. The reduction of the phenomenon of life to genetic information alone is perhaps reassuring because it is a somewhat familiar concept: it recalls the idea of an omnipotent 'word', abstract and almost metaphysical in nature, that would allow us to linearise the complex networks of natural phenomena by reducing them to a sort of instruction manual in which, we imagine, it may be possible to read our destiny clearly. But this genetic essentialism, in reducing life to the popular mythology of the 'selfish gene',[2] deprives it of its complexity

---

2.   The theory of the 'selfish gene' was formulated by the biologist Richard Dawkins in his 1976 essay of the same name. According to this position, living organisms are interpreted as simple 'vectors' of genetic information; life as a whole would therefore be a functional phenomenon collateral to the process of the self-replication of genes (R. Dawkins, *The Selfish Gene* [Oxford: Oxford University Press, 1976]).

and blinds us to the importance of the bottom-up processes of self-organisation which are the basis of all complex systems. As philosopher of science Evelyn Fox Keller wrote in her book *The Century of the Gene*,

> The image of genes as clear and distinct causal agents, constituting the basis of all aspects of organismic life, has become so deeply embedded in both popular and scientific thought that it will take far more than good intentions, diligence, or conceptual critique to dislodge it. So, too, the image of a genetic program—although of more recent vintage—has by now become equally embedded in our ways of thinking [...] I have argued throughout this book that our new understandings of the complexity of developmental dynamics have critically undermined the conceptual adequacy of genes as causes of development; furthermore, recent developments in molecular biology have given us new appreciation of the magnitude of the gap between genetic information and biological meaning.[3]

The first time a self-replicating molecular mechanism was observed in vitro was in 1967, when Sol Spiegelman and his collaborators isolated the RNA of the bacteriophage virus Qß and the corresponding enzyme that enabled it to replicate.[4] Spiegelman prepared a solution containing a certain amount of viral RNA and enzyme, along with the chemical nutrient used by the RNA for replication. Like any self-replicating molecule, RNA reproduces using an already-formed RNA molecule as a model: in this way, if a certain molecule is modified by a spontaneous mutation resulting from a copying error, this mutation will be transferred to subsequent 'generations'. After each replication cycle, a portion of the RNA and enzyme solution was transferred to a new nutrient solution for replication. The aim of the experiment was to isolate the self-replication process from all the other processes in the virus's 'life cycle': 'What will happen to the RNA molecules if the only demand

---

3.   E.F. Keller, *The Century of the Gene* (Cambridge, MA: Harvard University Press, 2002), 136.
4.   D.R. Mills, R.L. Peterson and S. Spiegelman, 'An Extracellular Darwinian Experiment with a Self-duplicating Nucleic Acid Molecule', *Proceedings of the National Academy of Sciences* 58 (1967), 217–24.

made on them is the Biblical injunction, *multiply*, with the biological proviso that they do so as rapidly as possible?'[5]

At the end of the 74th replication cycle, a chemical analysis of the molecules obtained revealed that the original RNA had degenerated into a much simpler molecule, destroying 83% of the information it initially contained and failing in its original function, i.e. becoming completely incapable of ensuring the production of viral particles. The experiment showed that a biological system reduced to its replication processes alone could not account for the complexity of living organisms, since the mere self-replication of information rewards speed of copying over complexity; this also called into question the possibility that life had emerged from a simple 'soup' of molecular replicators, demonstrating the importance of other *cooperative* physical processes in the birth of life. The 'de-facto' RNA molecule obtained from this in-vitro replication process was later renamed 'Spiegelmann's monster', with reference to Frankenstein as a predecessor.... In fact, despite their differences, the two stories do indeed seem to warn us against the same mistake. Like Frankenstein's monster in Shelley's tale, Spiegelmann's owes its monstrosity above all to its isolation from the relational fabric that allows life to fully realise itself, and constitutes its most authentic and profound meaning.

---

5.   Ibid., 217.

The Promises of Monsters

Advances in synthetic biology and in the production of forms of *lyfe* capable of embodying in artificial structures the same functions as those exhibited by living organisms have now consolidated the idea that life is not an exceptional phenomenon limited to biological matter, thus establishing a newfound continuity between living and non-living matter. Another consequence of this new understanding of life is that it is no longer possible to draw a rigid distinction between our technologies and living organisms. Just like Frankenstein's monster, our technologies are slowly and inexorably 'coming to life'. And it is principally thanks to chemistry, materials science, and nanotechnology that objects increasingly indistinguishable from living organisms—in composition, structure, and function—are beginning to see the light.

The idea of constructing a unified model of machines and living organisms can be traced back to 1948, when Norbert Wiener founded cybernetics, which he described as the 'science of communication and control in the animal and in the machine'.[1] According to Wiener, artificial machines and living organisms have in common the presence of similar feedback mechanisms, i.e. retroactive mechanisms capable of regulating and directing their behaviour spontaneously. Through these feedback mechanisms, machines and organisms attain a comparable level of autonomy: unlike rudimentary and primitive instruments that have to be directed from outside, the new breed of cybernetic machines are able to regulate themselves and need to be programmed only once and will then continue to carry out their tasks spontaneously. Cybernetics as a new paradigm of technology is applicable to the problem of artificial life because, at its most radical level, it problematises the

---

1.   N. Wiener, *Cybernetics: or Control and Communication in the Animal and the Machine* (Cambridge, MA: MIT Press, 2019).

relationship between human beings and their technological tools. While traditionally, artificial objects have no autonomy and depend entirely on human control to perform any function, 'cybernetic organisms' are, on the contrary, far more similar to us because they are able to function autonomously and to communicate with the outside world. Wiener defines the automaton as a necessarily intelligent and relational organism whose main function is to transfer information to and from the outside world:

> [T]he newer study of automata, whether in the metal or in the flesh, is a branch of communication engineering [...]. In such a theory, we deal with automata effectively coupled to the external world, not merely by their energy flow, their metabolism, but also by a flow of impressions, of incoming messages, and of the actions of outgoing messages.[2]

In 1970 Jacques Monod imagined an alien research probe visiting the earth to look for traces of life: without any previous knowledge of the specific characteristics of living organisms on earth, the alien, helped only by a computer, would be faced with the difficult task of distinguishing natural inorganic objects from living organisms, and living organisms from artificial objects built by humans or other animals.[3] The first task is quite easy: in order to distinguish living matter from non-living matter, the alien need only look for organised and regular structures, which manifest themselves almost exclusively in living matter; the only exceptions are inorganic crystals, which are however far simpler than any biological structure. Conversely, distinguishing artificial structures from natural ones is a far less obvious task: objects built by man, from neolithic arrowheads to the tyres of a modern car, as well as the structures produced by some animals, for example beehives, are endowed with a complex and regular structure which is repeated almost *identically* in many different objects; that is, they manifest a

---

2. Ibid., 60.

3. J. Monod, *Chance and Necessity: An Essay on the Natural Philosophy of Modern Biology*, tr. A. Wainhouse (New York: Knopf, 1971), 5–11.

THE PROMISES OF MONSTERS

property of *invariance* similar to that of living beings. What is more, the organs of living beings and the artefacts they construct seem to have as a common feature the fact that they have been designed to perform a specific function. In other words, unlike other natural objects, they are, according to Monod's definition, endowed with *teleonomy*. But according to Monod the imaginary alien's computer would ultimately be able to discover a fundamental difference between living organisms and artificial objects:

> The machine could not fail to note that the macroscopic structure of an artifact [...] results from the application to the materials constituting it of forces exterior to the object itself. Once complete, this macroscopic structure attests, not to inner forces of cohesion between atoms or molecules constituting its material (and conferring upon it only its general properties of density, hardness, ductility, etc.), but to the external forces that have shaped it.
>
> On the other hand, the program will have to register the fact that a living being's structure results from a totally different process, in that it owes almost nothing to the action of outside forces, but everything, from its overall shape down to its tiniest detail, to 'morphogenetic' interactions within the object itself.[4]

According to Monod, living beings are 'strange objects', and are also unique and unrepeatable in their strangeness, because their characteristics of *teleonomy*, *invariance*, and *spontaneous morphogenesis* are found in no other structure, either natural or artificial.

Let's repeat the same thought experiment in the light of the new technologies we have discussed above. Would Monod's alien be able to see the many artificial objects with complex structures, capable of spontaneous self-organisation, and in some cases self-replication, that our sciences have been able to design? Would it still be able to distinguish them from natural living organisms?

---

4.   Ibid., 10.

In the film *Terminator 2: Doomsday*, the young protagonist John Connor, a mere boy destined to lead human beings' rebellion against machines, finds himself fighting an almost indestructible cybernetic organism. Unlike its predecessors, the T-1000 is equipped with a completely liquid body that allows it to continuously change its shape: even when it seems to have disintegrated, it is capable of spontaneously reassembling itself and recovering its original structure. The mixed substance that makes up its body, a proteiform 'mimetic polyalloy' capable of imitating any object with which it comes into contact, is the ideal science-fiction example of the nanotech dream of an artificial organism whose shape is not determined by external forces, as with a traditional automaton, but is entirely encoded in the microscopic components that make it up. This new cybernetic Frankenstein not only imitates life, but even surpasses it in its capacity for self-organisation. While science is still a long way from designing a material with these kinds of capabilities, the design of artificial organisms capable of spontaneously assembling themselves is an increasingly realistic prospect, and challenges our ability to distinguish living organisms from artificial objects.

From *Frankenstein* to *Terminator*, the idea of violating the boundary between nature and technology by producing hybrid, ambiguous creatures generates a sense of deep anxiety. These science-fiction creatures are undoubtedly monstrous, especially in the sense that they cannot find a legitimate place in the natural order of things: they are neither living nor dead, neither natural nor artificial. The same sense of unease evoked by these fantastic narratives is also generated by all those synthetic organisms that call into question the rigid separation between living and non-living matter. From primitive Traube cells to the self-replicating synthetic molecules of contemporary chemistry, we may begin to wonder whether this desire to challenge the very definition of nature is rooted in an ambition of universal technological genius that is destined to backfire on us. Stories like Frankenstein, however, also suggest that the way we relate to these artificial 'monsters' retroactively affects the way they relate to us. How can we make these monstrous creatures valuable allies in building the technologies of the future?

This is the question that Donna Haraway asks in her 1992 essay 'The Promises of Monsters'.[5] According to Haraway, the objects of scientific investigation, such as life or nature, do not exist as realities independent of the science that studies them, but are the results of a *discursive process*. This does not mean that they are simply the product of an ideology and that they have no material basis, however. On the contrary, these categories are very real, but their definition, rather than being given a priori, is continually redefined by the scientific process, which does not merely 'discover' or 'reveal' a hidden reality, but rather *constructs* it in a bidirectional process in which, contrary to the idea of man's dominion over nature, nature in turn responds to human investigation and actively participates in its definition. According to Haraway, we must understand nature as an 'artifact',[6] i.e. a shared construction that arises in the encounter between human and non-human agents who participate in the creation of the categories that define them. In Haraway's words:

> Biology is a discourse, not the living world itself. But humans are not the only actors in the construction of the entities of any scientific discourse; machines (delegates that can produce surprises) and other partners (not 'pre- or extra-discursive objects', but partners) are active constructors of natural scientific objects . Like other scientific bodies, organisms are not ideological constructions. The whole point about discursive construction has been that it is not about ideology. Always radically historically specific, always lively, bodies have a different kind of specificity and effectivity; and so they invite a different kind of engagement and intervention.[7]

The project to generalise and modify the conventional definitions of nature and life does not aim to erase these concepts or to affirm that nature and technology exist in an indistinct continuity; rather, the aim of

5.   D. Haraway, 'Promises of Monsters: A Regenerative Politics for Inappropriate/d Others', in L. Grossberg et al. (eds.), *Cultural Studies* (London: Routledge, 1992), 295–337.

6.   Ibid., 332.

7.   Ibid., 298.

this cultural and scientific procedure is to build doorways and passages that allow us to create fruitful new *relationships* with the matter around us. This creative aspect of scientific investigation emerges over the long history of the encounter between chemistry and biology, in which, as we have seen, the definition of categories such as organic and inorganic, living and non-living, has continuously evolved, accompanied by the birth of new technologies capable of creating connections between different states of matter which, prior to this discursive process, were separated by seemingly insurmountable barriers. Today, the synthesis of organic molecules is the basis of the drugs we use to cure ourselves, and experiments in synthetic life allow us to penetrate the mystery of our origins while at the same time producing increasingly functional materials to meet the technological challenges of the future. In order to achieve this encounter between nature and technology, however, it is necessary to abandon the perspective of science as a unidirectional process that places humans in a position of domination over the objects they study and the tools they produce. 'Monstrosity', in this sense, is the result of the fruitful union between human and non-human agents, and is also the only way to build an authentic connection between nature and technology. The work of scientists such as Stéphane Leduc, promoters of a *synthetic approach* to the study of nature, is interesting for this very reason. In synthesis one cannot help but include one's object of study in the discursive process that defines it, making it an active part of the scientific investigation and allowing it to *speak* about its own identity. The transdisciplinary path of synthetic biology, running from the nineteenth century to the present day, began with the aim of studying the natural phenomenon of life; as it engaged more and more with the subject it studied, however, it no longer found its original object—the living being as traditionally understood—but a variety of new organisms emerging from the two-way interaction between nature and technology. Are these synthetic organisms still living? In asking itself this question, science has the opportunity to construct new epistemological categories that will allow it to free itself from hegemonic, sometimes oppressive, narratives about life and nature.

It is in this context that Haraway proposes the figure of the *cyborg*—which she defines as 'an implosion of the technical, organic, mythic, and political'[8]—as the new subject (rather than object) of techno-science. The cyborg is a child of the *interface*, the surface of exchange and communication between two different substances which, in chemistry, is also the crucial area where most reactions and interactions take place. As a cultural figure, the cyborg emerges out of a grey zone between deep, completely alien and inhuman space and the interior of the organism. The word 'cyborg', an abbreviation of the English expression *cybernetic organism*, was coined in 1960 by scientists Manfred Clynes and Nathan Kline in an article published in the journal *Astronautics*. According to these authors, in order to allow the human being to adapt to the extreme psychological and physical conditions imposed by space exploration, the body of future astronauts will no longer be augmented with external prostheses such as space suits, but will be modified from the inside, so that technological enhancements of the body and mind perfectly integrated with the biological organism.[9]

Although today we may be aware of the cyborg above all as the monstrous hybrid of science fiction stories, science has not abandoned the idea of an increasingly functional relationship between our bodies and new technologies. Nanotechnologies, and in particular advances in nanomedicine, offer new opportunities to interface biological organisms with artificial objects to build real hybrid organisms in which there is a vanishing distinction between the living body and non-living matter. This approach, as Clynes and Kline dreamed, could even be useful in space exploration in order to adapt the human organism to life in new inhospitable environments.[10] In more general terms, the cyborg is designed to modify its own body from within so that it can cross the boundaries of the human. In the cyborg, the connection between

8. Ibid., 300.

9. M.E. Clynes and N.S. Kline, 'Cyborgs and Space', in *Astronautics* (Orangeburg, NY: Rockland State Hospital, 1960), 26.

10. J. Patel and A. Patel, 'Nano Drug Delivery Systems for Space Applications', in Y. Pathak et al. (eds.), *Handbook of Space Pharmaceuticals* (Basel: Springer Nature Switzerland, 2019).

what is inside the organism and what is foreign to it, between what is human and what is inhuman, is achieved through a perfect integration of life and technology.

Reflecting on the relationship between the human and science in the light of the new discoveries of biology, Jacques Monod argued that in destroying any ambition for human centrality and questioning any universal value, science produced in man an incurable sense of deep dread and alienation. In *Chance and Necessity* he writes:

> The fear is the fear of sacrilege: of outrage to values. A wholly justified fear. It is perfectly true that science outrages values. Not directly, since science is no judge of them and must ignore them; but it subverts every one of the mythical or philosophical ontogenies upon which the animist tradition [...] has made all ethics rest: values, duties, rights, prohibitions.
>
> If he accepts this message—accepts all it contains—then man must at last wake out of his millenary dream; and in doing so, wake to his total solitude, his fundamental isolation.[11]

However, although it may be true that science forces us to renegotiate the human's central position in the universe and jeopardises time-honoured values, it can also act as a driver for new mythologies, such as that of the cyborg, which highlight the profound connection between human beings and matter around them. If the human can no longer be exalted as ruler of the cosmos, it can nevertheless rediscover a lost continuity with nature by accepting that there are no rigid barriers between inanimate matter, life, and technology. Guided by this new technological perspective, we awaken from the age-old dream of human domination over nature to immerse ourselves in new dreams in which, far from being alone, we find ourselves side-by-side with a multiplicity of unexpected and benevolent new forms of life. The problem that technology ought to pose to us, therefore, is not that of the legitimacy of crossing a sacred boundary, or that of defending some pre-established

---

11. Monod, *Chance and Necessity*, 172.

definition of nature. Science has always been a problem of synthesis, not of representation: it does not act as a mirror, reflecting distinctions that pre-exist in nature—such as the distinction between the living and the non-living—but always consists in the construction of new concepts and new organisms whose *monstrosity* allows us to navigate the complexity of reality. According to Haraway, this new vision of science implies 'unblinding ourselves from the sun-worshiping stories about the history of science and technology as paradigms of rationalism',[12] and recognising that the history of science is above all a cultural and discursive process of concept creation rather than a simple unveiling of nature. In the process, the renegotiation of the concepts of life and nature allows us to build new relationships with the materials around us, constructing a new continuity with our technologies and other living organisms. Rather than trying to consolidate the boundaries that divide the human from the inhuman, the natural from the unnatural, the living from the non-living, it is time for science to decide to forge a new alliance with its monsters.

12.   Haraway, 'The Promises of Monsters', 297.

5

WHAT THE FUTURE
IS MADE OF

Ariadne: *My king, my father, that's how heroes and gods see. What do you see during the day if not the night, fear, and the Minotaur that you have threaded from insomnia's film? Who cultivated his ferocity? Your dreams. Who brought him the first pack of young men and women, torn away from Athens by terror and the glory of sacrifice? He is your furtive creation, like the shadow of a tree is the vestige of a chilling night.*

Julio Cortázar, *The Kings*

Minds in the Web

The 'strange minds' we have explored, from *Physarum polycephalum* to octopi, from spider silk to artificial smart materials, call into question the conventional conception of the human mind as a centralised structure, organised around a single 'command centre' from which instructions for the control of the organism are sent out, and through which information about the external environment is collected. On the contrary, these minds owe their intelligence to their decentralised and diffused structure, in which a multiplicity of simple mechanisms combine to elaborate a complex response to the world in which they are located. It is therefore necessary to question, once again, the meaning of the word 'intelligence' when we use it to define non-human organisms and materials: it is now clear that although the mind, at least in our everyday perceptions, seems to function like a mirror, first building a unitary representation of reality and then acting within it, there are also minds that function in a *non-representative* way, without any need to build a reflective image of themselves and the world, and yet still manage to exhibit intelligent behaviour, adapting to their environment in response to external stimuli.

Diverse as they are, all of these minds are united by the fact that they consist of a very large number of elements which, considered individually, are 'stupid', i.e. do not exhibit the properties that the system as a whole manifests. The most important lesson to be learnt from these systems is that *intelligence emerges from relationships*: a set of simple interactions within a collectivity of elements can bring out properties that the individual components of the system alone did not possess. Intelligence may then be considered an emergent property of those systems we have learned to call complex systems, in which a multiplicity of parallel relations produce different forms of self-organisation from below.

The word *parallel*, in this context, serves to underline the fact that within these structures there is no pre-established hierarchical organisation: there is no 'control room' that directs the behaviour of the parts and there is no instruction manual that tells the individual components how to organise themselves with respect to one another. What is interesting about this vision of intelligence is that it does not depend on the specific nature of the components that constitute it, and for this reason it has been observed across disciplines, in many different areas: from quantum physics to biology, from psychology to information technology. In the case of materials science, the nanotechnological approach, focused on static and dynamic self-organising processes, makes it possible to exploit the emergent properties of complex systems to design new materials capable of behaving in an increasingly intelligent way. Spider silk is a perfect example of natural nanotechnology that exploits the complexity of its microscopic structure to optimise its interaction with the environment.

One might wonder why, if this vision of intelligence is so transversal and interdisciplinary, we should pay special attention to the existence of 'intelligent materials', a case that may seem marginal in an extraordinarily vast universe of other minds, natural and artificial. Many answers can be given to this question. As we have seen, on the one hand, studying the capacity of non-living matter to organise itself and receive stimuli from the outside world can help us find a continuity between the world of matter, which we usually consider inert, and the world of life, populated by dynamic structures capable of maintaining their own internal organisation, growing and replicating themselves. In discovering that the phenomena of self-organisation are not exclusive to life, we realise that our existence as living beings is not an exception, but just another confirmation of the vitality of matter that composes us, opening us up to the possibility of encountering forms of 'life' completely different from those that we know.

The search for alien life or the creation of completely artificial organisms may seem totally science-fictional propositions and, in the context of this essay, we have used them mainly as 'thought experiments' or 'proofs of concept' that help us to explore the outer reaches of an idea—that of the intelligence of materials—by taking it to its most radical consequences. But even if in most cases they are not (yet) able to completely assemble and replicate themselves, the materials that surround us in our daily reality do already participate very actively in the construction of the world in which we live. The materials that we use in our technologies, from the fabrics of our clothes to the silicon of our solar panels, from the cement of our houses to the Kevlar of the International Space Station, are an integral part of our culture, our history and, above all, our future. For this reason, the position we decide to take with regard to the materials that make up our world is of fundamental importance. Our intelligence and the intelligence of the materials we use are connected, networked, and exert influence on one another to determine the shape of our reality. Is there a way to think of the materials that surround us not as passive objects, but as an active part of our world?

An answer to this question may be found in the extraordinary material from which we started and in its relationship with the mind of the animal that produces it. In a 2017 article, biologists Hilton Japyassù and Kevin Laland proposed that the spider is one of the very few animals, apart from humans, capable of extending its mind outside of its body.[1] In Chapter 2 we encountered the use of the concept of extended cognition, or the extended mind, to describe humans' ability to 'delegate' a part of their cognitive work to the objects around them, for example using computers and smartphones to communicate with our fellow humans, calculators to facilitate arithmetic calculations, or notebooks as an extension for long-term memory. Similarly, Japyassù and Laland propose that the spider uses the web it builds as an extension of its perceptual organs and its brain. According to them, since the spider is a predator living in a very complex environment, and since maintaining a highly developed central nervous system has an extremely high energy cost, evolution has 'chosen' to transfer part of the spider's intelligence, which the animal needs to hunt and orient itself in space, to a material external to its body.

1.   H.F. Japyassù and K.N. Laland, 'Extended Spider Cognition', *Animal Cognition* 20 (2017), 375–95.

The spider is almost completely blind and has a rather simple central nervous system, which makes it incapable of storing long-term information or constructing a mental representation of its surroundings. In spite of this, it is able to orient itself within the complex three-dimensional space it inhabits, building with its own silk perfectly symmetrical structures that are of enormous dimensions relative to its own body, something that would be very demanding even for a human individual. The way in which the spider manages to accomplish such a complex task is determined precisely by its ability to use silk to draw a geometric map of the space around it, using it as a sort of spatial memory external to its body. We have already encountered a similar behaviour, although with significant differences, in polycephalic slime's ability to map and explore its environment by depositing chemical traces of its passage which act as a memory external to the body. According to Japyassù and Laland, 'the spider, by relying on the previous threads as external, long-term memory devices, probably requires less CNS long-term memory than other similarly complex animal activities'.[2] In this sense, it becomes entirely impossible to locate the boundary between the spider's mind and its web.

But the cognitive relationship between spider and web is not limited to determining the animal's orientation in space. The web acts as a genuine sense organ for the spider, transmitting and amplifying the vibrations of the prey trapped in its threads. This capacity for transmission is closely related to the microscopic structure of the silk, its thickness, and the tension given by the spider to the threads of the web. The authors report that hungry spiders increase the tension in their webs to increase their sensitivity, allowing them to respond to smaller prey that they would usually ignore. Not only that, but by artificially altering the tension in certain areas of the web, the experimenters were able to alter the spider's mind, focusing its attention on the most sensitive areas. 'In this sense', they write, 'web threads cannot be understood as passive transmitters, or even passive filters of vibratory information. Thread properties are adjustable and thus can process the same information in

---

2.   Ibid., 385.

different and adaptive ways.'[3]

From this perspective it is clear that the relationship between the spider and its silk is a very tightly-woven one. On the one hand, the spider modifies its perception of the world by altering the canvas both from an engineering point of view (i.e. modifying its shape and tension), and a nanotechnological point of view, improving and adapting its microscopic chemical structure during evolution to make it more functional for its own purposes. On the other hand, the silk itself, as a complex and sensitive material, influences the way the spider perceives the world and acts upon its environment, to the point where the smallest mutation in the chemical structure of silk can bring about a radical change in the spider's behaviour. It is precisely this mutual influence between *animal mind* and *material mind* that allows the spider to succeed in its environment; to think of silk simply as a passive object which the spider uses to survive is an oversimplification of this complex reciprocal relationship.

Our cultural relationship with the materials we use in many ways resembles that between the spider and its silk. The web the spider builds is both an artificial technology and a natural environment, both a tool and an integral part of the subject that produces it. We too are defined as human beings by our ability to shape our environment using the materials around us, which in turn influence our reality and our future. We might perhaps imagine that our relationship with the materials we use is less intimate: after all, the structure of silk is written in the spider's genes, while our materials seem more separate from us: it is, however, increasingly evident that the future of our species is also inextricably intertwined with the materials we decide to use to build it. In this sense, the concept of the *web* or *network* may be used as a transdisciplinary category to address our relationship with the materials of the future. From a technological point of view, designing materials with a relational internal structure, i.e. a network of numerous components that participate in a fabric of parallel and reciprocal connections, allows us to build technologies that are increasingly intelligent and sensitive, like

---

3.   Ibid., 381.

spider silk. At the same time, from a cultural point of view, replacing a vertical vision of our relationship with technology in which human beings univocally define the relationship with their own instruments, with the idea of a network of parallel interactions between human and non-human agents, in which the materials we use can in turn influence our relationship with the world, allows us to extend our mind and open ourselves up to a more ecological vision of our relationship with the environment around us.

Arachne 2.0

One of the most interesting issues raised by the study and synthesis of new materials concerns the relationship between nature and technology. As we have discovered, many new artificial materials are characterised by the ability to enter into relation with living organisms, to mimic the characteristics of natural materials, and to assemble into hybrid objects in which nature and technology blur and blend together. Here again we can use spider silk as a 'model material' to reflect on the hybridisation between nature and technology. In itself, the spider's ability to weave its own web could be understood simply as a natural fact and put down to the difficult-to-define idea of *instinct*; but at the same time, the spider's web is the technological tool the creature uses to modify the environment around it.

Since the birth of nanotechnology, carbon has been the most important element for the synthesis of new materials on the nanoscale. Graphene and related materials, such as carbon nanotubes, owing to their mechanical, thermal, and electrical properties, are now considered the nanomaterials par excellence. Graphene is a two-dimensional material made up of a graphite 'sheet' in which carbon atoms are arranged to form hexagonal cells. Nanotubes have the same chemical structure as graphene, but are 'rolled up' on themselves to form straw-like tubes. The interaction between these artificial materials and spider silk, owing to their very high surface area, produces an extraordinary reinforcement, generating what in materials science is called a *nano-composite*: a dispersion of nanometric particles, usually inorganic, in a matrix, usually organic. Nanomaterials such as graphene are often not used alone, but find their principal applications in the preparation of such nanocomposites, where nanoparticles of the material are combined with other substances to form hybrid materials with improved or completely new properties. A study in 2017 analysed the effect on the physical characteristics of spider webs when they are exposed to certain artificial nanomaterials,[1] and showed that, after ingestion of graphene and carbon nanotubes, spiders are able to weave a silk up to ten times more tenacious, i.e. ten times more capable of absorbing the energy of an impact, and three times more resistant than ordinary silk. The material obtained from this process yields a fibre with some of the most extreme mechanical properties ever recorded, surpassing even the most advanced synthetic fibres. This nanotechnological silk is one of the first and most fascinating examples of a *bionic material*, i.e. a hybrid material obtained by combining a natural material produced by a living organism with an artificial material. Here the spider's unparalleled ability to weave its web, so difficult to reproduce in the laboratory, is integrated with some of the most innovative man-made materials. Spider silk modified with artificial nanomaterials could also be used for electronic and robotics applications, where its resistance, combined with the electrical conductivity imparted by the nanoparticles dispersed within it, could allow us to produce circuits, sensors, and tiny artificial muscles. In fact, by exploiting the spider web's ability to contract and extend as the ambient humidity varies, it is possible to use the electrical resistance of the bionic silk thread to heat and cool it in a controlled way, obtaining contractions whose capacity to perform mechanical work, if considered in proportion to the very fine diameter of the fibre, would far exceed that of human muscles.[2]

The idea of using innovative materials capable of forming an extensive and dynamic interface with biological structures is attracting growing interest. Because of their scale and properties, these carbon-based nanomaterials transform nature as we know it, even potentially becoming part of our own bodies. One example of this concept is a newly designed prosthetic material for spinal lesions. A 'sponge' made up

1.   E. Lepore, et al., 'Spider Silk Reinforced by Graphene or Carbon Nanotubes', *2D Materials* 4:031013 (2017).

2.   The construction of such a device, obtained from spider silk threads covered with carbon nanotubes, was reported in the magazine *Nature Communications* in 2013 (E. Steven et al., 'Carbon Nanotubes on a Spider Silk Scaffold', *Nature Communications* 4:2435 [2013]).

of carbon nanotubes, when introduced inside a damaged area of the spine, could be an effective support for the regeneration of neurons: not only is it structurally resistant, owing to the mechanical properties of carbon nanotubes, but it also integrates the functionality of nerve cells owing to its outstanding electrical conductivity.[3] The carbon nanotubes used are produced following a self-assembly approach which exploits the ability of carbon atoms to self-organise into precise structures out of a chemical hydrocarbon vapour. Here the intelligence of inorganic matter intertwines with living intelligence in an extraordinary hybridisation of life and chemistry that could potentially find a role to play in even the most intimate mechanisms of our own nervous system. These examples highlight one of the most interesting aspects of nanotechnology, namely the possibility of acting on the same level as biological life. No doubt this also presents risks, many of which will need to be clarified before we can feel comfortable using these as wide-scale technologies: owing to their very small size, nanoparticles can penetrate undisturbed into our bodies, where, depending on their chemical nature and structure, they could cause irreversible damage; the difficulty of filtering and recovering nanoparticles once they have been dispersed into the environment makes the problem of using nanotechnologies yet more complex.[4]

From a more conceptual point of view, materials science brings life and technology onto an equal footing, thus affording us the opportunity to renegotiate our relationship with the technologies we build. The question of where nature ends and technology begins may perhaps seem trivial, but the ever-expanding interface between new materials and our bodies shows that it is precisely on the borderline between these two worlds that the most interesting technological potentials are realised, and it is in this hybrid space that applied science, so often excluded from the more theoretical discussions on the role played by

---

3.   S. Usmani et al., 'Functional Rewiring Across Spinal Injuries via Biomimetic Nanofiber Shelves', *Proceedings of the National Academy of Sciences* 117:41 (September 2020), 25212–18.
4.   An overview of the potential risks of nanoparticles for human health can be found, for example, in M.M. Sufian et al., 'Safety Issues Associated with the Use of Nanoparticles in Human Body', *Photodiagnosis and Photodynamic Therapy* 19 (2017), 67–72.

science in our world view, can and must also open onto an epistemological and political dimension. Undoubtedly, the development of science and technology has often provided the pretext for bolstering the cultural paradigm that places humans in a position of domination over the living and non-living matter that surrounds them, with often catastrophic consequences. This domination is rooted in the possibility of drawing insuperable boundaries between what makes us human and what is radically different from us. On the other hand, any appeal to the concept of nature, whether to protect it or to exercise dominion over it, ends up reaffirming the distance that separates us from it, relieving us of the responsibility of questioning our position in the world and our use of technology. In contrast to this approach, the construction of hybrid technologies that transcend the presumed boundaries of what is natural and what is technological can help us establish a fruitful dialogue with the objects and organisms with which we share our culture, calling into question the idea that there is only one legitimate view of the universe we inhabit.

Like spiders in their webs, our knowledge of reality is shaped by the tools and materials we use to relate to the environment. If the mind of the spider extends into the depths of the microscopic structure of its silk fibres, what does the spider-cyborg dream of, asleep in its bionic graphene-wired web? What unknown prey's vibrations reach the neurons in its eight legs? How will information flow through our spine when its marrow is intertwined with carbon nanotubes? Such strange hybridisations reveal the transience of the boundary that separates our consciousness from the world around us, and our objects of study from the subject that studies them. By creating completely new tools and environments, the technology of new materials does not just allow us to discover a pre-existing nature that reveals its secrets to us; it actively contributes to the construction of new realities. Since our mind, like that of the spider, is inextricably intertwined with our technologies, building new materials means above all inventing new ways of relating to the world, new ways of seeing and feeling the matter around us, produced through the encounter between our intelligence and other minds.

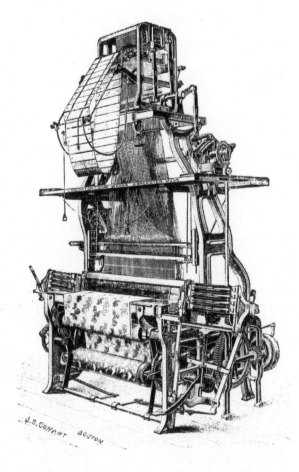

Jacquard Loom, illustration from *The Popular Science Monthly*, 1891.

Weavers of the Future

Arachne, the mythical weaver whose story we told in the opening pages of this book, embodies the idea of a technology that intertwines perfectly with she who produces it. Not only does Arachne weave the threads of her tapestry with incredible skill, producing complex structures from the mutual interaction of simple elements, she also transforms herself into a spider, blending herself and her loom into a single hybrid body that incarnates the indissoluble weave between mind, body, technology, and nature. Sigmund Freud, in his *Introduction to Psychoanalysis* in 1932, suggests that women, despite their general inability to participate actively in the development of human civilisation, have made at least one contribution to the history of technology:

> It seems that women have made few contributions to the discoveries and inventions in the history of civilization; there is, however, one technique which they may have invented—that of plaiting and weaving. If that is so, we should be tempted to guess the unconscious motive for the achievement. Nature herself would seem to have given the model which this achievement imitates by causing the growth at maturity of the pubic hair that conceals the genitals. The step that remained to be taken lay in making the threads adhere to one another, while on the body they stick into the skin and are only matted together.[1]

---

1.   S. Freud. 'New Introductory Lectures on Psychoanalysis. Lecture 33: Femininity', in J. Strachey (ed.), *The Standard Edition of the Complete Psychological Works of Sigmund Freud* (London: The Hogarth Press and the Institute of Psycho-Analysis, 24 vols, 1943–1974), vol. 22, 112–135: 132.

According to Freud's vision, the ability to weave is inherent in women's nature, because women learned weaving by imitating the natural weave of hair which, according to the psychoanalyst, 'hides' the 'lack' of the penis that every woman unconsciously desires. Although, on the one hand, this is one of the most misogynistic pages in the history of psychoanalysis, in which all contribution to technology and science is completely denied in the name of an inscrutable biological destiny, Freud is nonetheless correct in at least one respect: that weaving and femininity are connected in a deeper way than it may seem. Not only has weaving made an incalculable contribution to the development of our culture, it has more or less explicitly been a model for many of the most advanced technologies we use today, from artificial intelligence to nanotechnology. Addressing the relationship between cybernetics, computing, and femininity, Sadie Plant focuses on the central yet often neglected role played by weaving in the birth of the first computers and the development of technology as a whole. As she states:

> Perhaps weaving is even the fabric of every other discovery and invention, perhaps the beginning and the end of their history. The loom is a fatal innovation, which weaves its way from squared paper to the data net.[2]

To which we might add that, as we have seen, weaving is an art that begins with the loom and ends up intertwined with the intelligence of our most advanced nanomaterials. While this does not, of course, mean adhering to the idea that women possess some kind of exclusive capacity rooted, as Freud argued, in the female psyche, it does highlight the need to modify our cultural approach to technology.

Many contemporary feminist theorists have questioned the relationship between nature and culture, reflecting on the possibility of negotiating a new relationship between science, the matter it studies,

---

2.   S. Plant, 'The Future Looms: Weaving Women and Cybernetics', *Body and Society* 1:3–4 (1995). In this text, Plant refers to the influence of the invention of the Jacquard loom on the development of the first computational machines, in particular the Analytical Engine designed by Ada Lovelace and Charles Babbage in the 1840s.

and the technologies it builds. The reason why femininity and technology are so often considered incompatible in our patriarchal civilisation is rooted in the idea that science and technology are tools that man uses to exercise a form of violent domination over nature. From this cultural perspective, an active subject, the man-scientist, acts upon a passive object, matter, giving it form and bending it to his own will. If we were to exemplify this approach in one of the most primitive human technologies, the process of splintering a stone, transforming it into an axe or arrowhead, is a good embodiment of the idea that matter is an essentially inert object which does not work with, but opposes, our attempts to modify it to our advantage. This paradigm of man's dominion over matter, which places him in the privileged position of having to give shape to an essentially blind and stupid substance, then also exerts its effects on the social and political level: everything that is perceived as 'other'—that is, in some way, 'less human'—becomes an object of domination and violence. Is it possible to imagine a technological paradigm in which matter actively participates in its own process of transformation?

According to the philosopher Luce Irigaray, modern Western metaphysics is based on the idea of a universe made up of solid bodies in rigid interaction with each other. This vision is reflected in the idea of the interaction between bodies as a collision or clash rather than a relationship: the only thing these rigid objects can do is bump into each other, a bit like the process of chipping a stone to turn it into a tool. On the contrary, Irigaray argues that the substratum of reality is essentially fluid, i.e. made up of bodies without rigid boundaries, which interpenetrate and mix with one another.[3] As in the weave of a fabric, individuals intertwine with one another to produce a network

---

3. Irigaray develops this critique in the series of essays *Marine Lover of Friedrich Nietzsche*, tr. G.C. Gill (New York: Columbia University Press, 1991), *Elemental Passions*, tr. J. Collie and J. Still (London: Routledge, 1999), and *The Forgetting of Air in Martin Heidegger*, tr. M.B. Mader (Austin, TX: University of Texas Press, 1999), in which she argues that the forgetfulness of fluid lies at the basis of Western metaphysics. For a detailed exposition of Irigaray's thinking on physics see J. Bardsley, 'Fluid Histories: Luce Irigaray, Michel Serres and the Ages of Water', *philoSOPHIA* 8:2 (2018), 13–38.

in which individual boundaries blur and individuals become indistinguishable. This continuous interpenetration can serve as the basis of an understanding of the material environment in which we are immersed and the other human beings around us as part of a single network of relationships, one which for this reason also implies a shared responsibility to construct an ethics based on the recognition of the other. In this sense, weaving embodies a relational and feminist vision of technology as an indissoluble interweaving of human, animal, and material agents.

While exploring the boundary between organic and inorganic we have already met the figure of the cyborg, the cybernetic organism which, according to Donna Haraway, could constitute a new paradigm for technology, capable of overcoming the binary logic that separates nature and culture. After all, our journey through the world of new intelligent materials has brought us to the confrontation with many 'monsters', some mythological and some technological, some natural and others artificial: from the Lernaean Hydra to the Golem, from polycephalic slime to bionic spiders, from Frankenstein's creature to self-replicating organic molecules. What all of these strange organisms have in common is the challenge they pose to our customary notions of what belongs to the natural order of things; they inhabit the borderland between life and death, mind and body, technology and nature.

'It is not just that science and technology are possible means of great human satisfaction, as well as a matrix of complex dominations', writes Haraway in her famous *Cyborg Manifesto*. 'Cyborg imagery can suggest a way out of the maze of dualisms in which we have explained our bodies and our tools to ourselves'.[4] Haraway's scientific training, which led her to address the concept of organism in the history of biology, naturally leads her to the conclusion that there is 'There is no fundamental, ontological separation in our formal knowledge of machine and organism, of technical and organic'.[5] In fact, as we have seen, all of these categories are extremely fluid, and can merge into each other with

---

4.   D. Haraway, 'A Cyborg Manifesto: Science, Technology, and Socialist-Feminism in the Late Twentieth Century', in *Simians, Cyborgs and Women: The Reinvention of Nature* (New York; Routledge, 1991), 149–81: 181.

5.   Ibid., 178.

great ease. From Haraway's radical perspective, while the increasingly pervasive hybridisation of technology and life might frighten us, and is often rejected by more traditional feminisms as an expression of man's technological domination over nature, this prospect actually harbours potential for emancipation:

> That is why cyborg politics insist on noise and advocate pollution, rejoicing in the illegitimate fusions of animal and machine. These are the couplings which make Man and Woman so problematic, subverting the structure of desire, the force imagined to generate language and gender, and so subverting the structure and modes of reproduction of 'Western' identity, of nature and culture, of mirror and eye, slave and master, body and mind.[6]

The link between a certain strand of feminist political thought and technology passes via a renegotiation of the scientific and cognitive gaze upon nature. This does not mean, however, that technology allows us to shape reality at will; rather, it means that our experience of the world is always intertwined with our tools and the materials we use, which form a dense and inextricable weave with our minds. In other words, our scientific knowledge of reality is not a more or less perfect mirroring of a world of passive and distant objects. On the contrary, knowledge of reality, if not reality itself, is produced in the *encounter* and the *relationship* between ourselves and our object of study. This is the perspective adopted by Karen Barad, who, starting from an analysis of the role of the measurement process in quantum physics, arrives at a new definition of the human relationship with matter.[7] In the background of Barad's thought is the essential and most well-known problem of quantum mechanics, the problem of indeterminacy. Without entering into the technical details of this problem, whose consequences have now become common knowledge, the principle of indeterminacy implies that in the study of quantum objects it is impossible to separate

---

6. Ibid., 176.

7. K. Barad, *Meeting the Universe Halfway: Quantum Physics and the Entanglement of Matter and Meaning* (Durham, NC and London: Duke University Press, 2007).

the process of measurement via which we *know* the object we are study-ing, from the *physical nature* and *properties* of the object itself. In other words, quantum objects seem to behave differently, typically as waves or particles, depending on *how we look at them*, i.e. depending on the instrumental apparatus with which we choose to study them.

In the thinking of the physicist Niels Bohr, one of the fathers of quantum mechanics, this problem is solved by resorting to the concept of complementarity, according to which the behaviour of the studied quantum object cannot be separated from the measuring apparatus that studies it. Setting out from this idea, Barad proposes a new vision of the physical universe as no longer constituted by single objects that pre-exist their scientific investigation, but rather as a network of rela-tions within which physical phenomena take shape and acquire mean-ing. According to Barad, the concept of matter 'refers to the materiality and materialisation of phenomena, not to an assumed, inherent, fixed property of abstract, independently existing objects':[8] matter is the process that results from an encounter rather than an a priori existing reality. In this vision and process of matter, the object being studied and the mind of the scientist who studies it participate symmetrically and cooperatively in the definition of knowledge. The concept of *entan-glement*, which in quantum physics indicates the indissoluble correla-tion between two particles of the same system, is repurposed by Barad as a more general paradigm of the human relationship with matter, a relationship in which there are no defined boundaries but only a con-tinuous and reciprocal influence. This perspective on science then pre-sents an opportunity to question the rigid separation between subject and object that runs through the history of modern Western thought:

> There is no *res cogitans* that inhabits a given body with inherent boundaries differentiating self and other. Rather, subjects are dif-ferentially constituted through specific intra-actions. The subjects so constituted may range across some of the presumed boundaries (such as those between human and nonhuman and self and other)

---

8. Ibid., 210

that get taken for granted. Knowing is a distributed practice that includes the larger material arrangement. To the extent that humans participate in scientific or other practices of knowing, they do so as part of the larger material configuration of the world and its ongoing open-ended articulation.[9]

Abandoning the strange world of quantum particles to return to the familiarity of our spider webs, the relational vision of matter that Barad applies to theoretical physics can very easily be extended to the applied sciences and, in the light of what we have seen above, lends itself in particular to the case of intelligent materials. For in this context also, we are dealing with objects which, although very different from Bohr's quantum particles, *actively participate in the construction of the reality in which they are immersed*. In particular, we have already highlighted how chemistry and materials science propose an essentially synthetic approach to the study of matter, in which the product of the cognitive process is always also the product of a creative and productive process which leads to the birth of new bodies. The synthesis of a new material is never the result of a univocal process determined by the human experimenter, because it always exploits the ability of matter to organise itself spontaneously, bottom-up. The challenge of synthesis, then, is not so much to dominate matter, but rather to understand 'what a material can do'—that is, to reveal its intrinsic vitality and its deepest intelligence, as manifested in its relationship with us.

Among the authors who have embraced this relational vision of matter and technology is the philosopher Jane Bennett who, in her book *Vibrant Matter: A Political Ecology of Things*, proposes what she calls a 'vital materialism'. According to Bennett, the inorganic materials and bodies around us are also endowed with an intrinsic vitality, which is manifested in their ability to actively participate in our world. From the perspective of this author, it is not possible to imagine a material environment separate from our human culture: on the contrary, culture and environment are connected on a single horizontal ontological plane

---

9.   Ibid., 379.

without pre-established hierarchies, and every agent participating in it, from humans to animals to the objects we use in our daily lives, is capable of actively communicating with every other agent.[10] To describe this profound connection, rather than using the term 'matter', a word that in philosophical language has taken on the meaning of a passive object rigidly separated from the human subject, Bennett prefers to speak of *material configurations*:

> I am a material configuration, the pigeons in the park are material compositions, the viruses, parasites, and heavy metals in my flesh and in pigeon flesh are materialities, as are neurochemicals, hurricane winds, *E. coli*, and the dust on the floor. Materiality is a rubric that tends to horizontalize the relations between humans, biota, and abiota. It draws human attention sideways, away from an ontologically ranked Great Chain of Being and toward a greater appreciation of the complex entanglements of humans and nonhumans. Here. the implicit moral imperative of Western thought—'Thou shall identify and defend what is special about Man'—loses some of its salience.[11]

By placing the relational fabric of matter at the centre of its ontology, this new feminist and materialist thinking invites us to enter into a direct relationship with the technologies that make up our world, to understand their impact in a deeper way and, if necessary, to take greater responsibility for the network of which we are a part. From my point of view, it is very significant that authors such as Haraway and Barad, whose influence and popularity are ever-growing and who have been promoters of some of the most radical and timely visions of our relationship with technology, are not only philosophers, but also women and scientists. Their scientific experience in the field is evident in the way in which they treat science and technology not as mere theoretical

---

10. J. Bennett, *Vibrant Matter: A Political Ecology of Things* (Durham, NC and London: Duke University Press, 2010), 116.

11. Ibid., 112.

expedients, but as real interlocutors in the elaboration of their own thought. In their work the countless configurations that matter can enter into, and its intrinsic intelligence and vitality, become a starting point for radically rethinking our position in the world.

Fifteenth-century engraving of the labyrinth and the story of Theseus and Ariadne. Image: © The Trustees of the British Museum

Ariadne's Thread

It is undeniable that, from the moment we started using technology—that is, probably long before we became human—we entered into a labyrinth whose complexity becomes greater and greater the deeper we reach. I like to imagine the path we have walked together in these pages as a passage through some corridor within this labyrinth: not a linear tale of progress, but a winding road full of surprises, blind alleys, mistakes, and unexpected discoveries. After all, no science, let alone a young and applied science such as materials science or nanotechnology, has a simple story to tell; often, those who decide to write about science are forced to choose what to include and what to exclude, finding themselves spinning a yarn even when there was not one there before.

The labyrinth is populated by monsters: hybrid creatures that we produced more or less consciously in our wonderful and unexpected encounters with matter. Some of these monsters, like those I have introduced within these pages, offer us the opportunity to forge new and fruitful alliances with the matter around us, positively modifying our environment and our culture. Others, far more threatening, jeopardise the possibility of our building a future for our species.

Daedalus, the mythical architect of the labyrinth of Knossos, designed a maze so intricate and perfect that, when he was locked within its walls with his son Icarus by the tyrant Minos, he himself, although he was the creator, was unable to find his way out. Often our destiny seems to resemble that of Daedalus: technological progress has led us to produce instruments so complex and extraordinary that we risk being trapped within the labyrinth produced by their consequences. Today it is hardly possible to speak of science and technology without underlining that the climate catastrophe now devastating our planet is the direct consequence of our actions, and is the fruit of an obsolete technological paradigm, that of fossil fuel consumption, which should have been entirely replaced decades ago with alternative technological solutions. I have chosen not to focus here on a detailed description of the technologies that materials science can offer to try and help stem the climate catastrophe, but they are many: the intelligence of new materials can be placed in the service of a more sustainable future by limiting our emissions of CO_2 and radically changing the way we produce, store, and consume energy.[1]

Faced with the ever-increasing potential of our technologies, it is easy to succumb to the temptation to take up an entirely techno-optimistic position. Unfortunately, although new technologies for energy production and storage can be further disseminated and improved, their ability to really change things is inextricably linked to our cultural and political fabric. Many of the materials that promise to bring about a 'green revolution' contain rare elements, non-renewable resources whose extraction has a social and environmental impact potentially comparable to that of oil. The effects of the reckless use of nanotechnology are still unknown and largely uncontrollable, precisely because of the ability of nanomaterials to interact with the most intimate structures of living organisms.

In any case, it is the relationship of domination that humans have established over the surrounding environment that is unsustainable: the search for alternative energy solutions does not touch the root of the problem, which is our inability to understand ourselves as a single node in the wider network of organisms and materials with which we share our world. If the monsters described in this book had not sufficed, the close encounter with a virus which, over the course of a few months, completely upended our way of life—a hybrid object, natural but at the same time artificial, a result of the exponential anthropisation of the planet—should have been enough to convince us that it is not necessary for an organism to be equipped with a brain, nor even to

---

1.   For an informative exhibition of how new materials can change the future of energy, I recommend Luca Beverina's essay *Futuro Materiale. Elettronica da mangiare, plastica biodegradabile, l'energia dove meno te l'aspetti* (Bologna: Il Mulino, 2020).

be alive in the strict sense of the word, for it to be able to intertwine its destiny with our own.

Although we undoubtedly have a responsibility to improve the situation by building more sustainable materials and more environmentally friendly technologies, imagining that some new technology or the dawn of a new science will come to save us—whether we are talking about geoengineering, biotechnology or nanomaterials—is perhaps just to naively look for an easy way out from the labyrinth in which we are trapped. We know very well how Icarus's attempt to use wax wings to fly out from between the walls of his father's labyrinth ended; looking for a technological solution to lift us above the complexity of our situation may be delude us for a while that we are in control, but it risks ending in ruin. But Icarus's flight certainly has something to teach us. Our desire to transcend the network that connects us to our technologies and to the other organisms that inhabit our planet, our claim to look at the world from a privileged perspective, will not help us solve the challenges that lie ahead; on the contrary, it will precipitate us straight onto the horns of the monsters we are fleeing. The reality is that there are no simple paths, and never will be. The way out of the labyrinth will never be upwards, but always horizontally, following its twisted ways.

The young Ariadne, perhaps because she is a woman, an expert like all Greek maidens in the ancient female art of weaving, comes up with a different solution. To her beloved Theseus, who is about to plunge into the depths of the labyrinth to defeat the monster that roams its corridors, Ariadne gives a ball of thread. Perhaps a rather humble technology compared to the superior engineering of Icarus's wings or Theseus's glittering bronze sword, Ariadne's thread nonetheless harbours a subtle form of intelligence: bending and adapting to the twisted curves of the labyrinth, flowing silently in parallel to its walls, it intertwines with the mind of the human being who explores it, allowing him to remember his path. This ancient tale should remind us that the only way out of the problem of our relationship with technology is through it: the technologies that will guide us out of the labyrinth will have to be more like Ariadne's thread than Icarus's wings, perhaps more humble and less ambitious, but cunning and flexible, capable of relating to the complexity of the reality in which we live. In this journey, the intelligence of the materials we use to orient ourselves in the world must interweave inextricably with our own, like the threads of a fabric.

Acknowledgements

I would like to express my heartfelt gratitude to my research group at Bicocca University, NanoMat@Lab, in particular Barbara Di Credico, Massimiliano D'Arienzo and Roberto Scotti, for having offered me the wonderful opportunity to do my doctorate in nanomaterials.

I would like to thank Enrico Monacelli for the precious comparison with philosophy of mind, in particular in relation to John Searle's thought, as covered in Chapter 2. I owe to Louis Fabrice Tshimanga the discovery of the beautiful manual by Bertuglia and Vaio, *Complessità e Modelli*, which was a fundamental guide for the writing of the chapter 'Structure and Function'. Many thanks to Matteo De Giuli for his support for this project and my writing activity in general.

I owe to Holly Heuser a deeper, and not just scientific, understanding of what it means to build and inhabit a network. And it was thanks to Matteo Grilli that I did not abandon this project halfway through these complex months.

Simone Sauza and Elena Tripaldi made an invaluable contribution to the revision of the drafts. I thank them from the bottom of my heart for their care, intelligence, and closeness. Thanks to my parents Elisabetta Binaghi and Fabio Tripaldi: the interdisciplinary nature of this essay is an attempt to repay the enthusiasm they have always transmitted to me.

Bibliography

Adamatzky, Andrew et al. 'Physarum Imitates Exploration and Colonisation of Planets', in *Advances in Physarum Machines: Sensing and Computing with Slime Mould,* ed. A. Adamatzky. Basel: Springer, 2016.

Álvarez-González, Begoña et al. 'Cytoskeletal Mechanics Regulating Amoeboid Cell Locomotion'. *Applied Mechanics Reviews* 66:5 (5 June 2014), 0508041–05080414.

Awad, Abubakr Hussien et al. 'Physarum Polycephalum Intelligent Foraging Behaviour and Applications—Short Review'. (2 March 2021). Preprint, arXiv:2103.00172v1.

Barad, Karen. *Meeting the Universe Halfway: Quantum Physics and the Entanglement of Matter and Meaning.* Durham, NC and London: Duke University Press, 2007.

———, 'Nature's Queer Performativity'. *Qui Parle* 19:2 (1 December 2011), 121–158.

Bardsley, Jessica. 'Fluid Histories: Luce Irigaray, Michel Serres and the Ages of Water'. *philoSOPHIA* 8:2 (12 August 2018), 13–38.

Barkan, Terrance. 'Graphene: The Hype versus Commercial Reality'. *Nature Nanotechnology* 14 (3 October 2019), 904–906.

Bartlett, Stuart and Wong, Michael L. 'Defining Lyfe in the Universe: From Three Privileged Functions to Four Pillars'. *Life* 10:42 (16 April 2020).

Bateson, Gregory. *Mind and Nature: A Necessary Unity.* New York: Dutton, 1979.

Bengisu, Murat and Ferrara, Marinella (eds.). *Materials that Move: Smart Materials, Intelligent Design.* Basel: Springer, 2018.

Bennett, Jane. *Vibrant Matter: A Political Ecology of Things.* Durham, NC and London: Duke University Press, 2010.

Beverina, Luca. *Futuro Materiale. Elettronica da mangiare, plastica biodegradabile, l'energia dove meno te l'aspetti.* Bologna: Il Mulino, 2020.

Brunetta, Leslie and Craig, Catherine L. *Spider Silk: Evolution and 400 Million Years of Spinning, Waiting, Snagging, and Mating.* New Haven, CT: Yale University Press, 2010.

Carnall, Jacqui M. A. et al. 'Mechanosensitive Self-Replication Driven by Self-Organization'. *Science* 327:5972 (19 March 2010), 1502–1506.

Clark, Andy and Chalmers, David J. 'The Extended Mind', *Analysis* 58:1 (January 1998), 7–19.

Clark, Andy. *Being There: Putting Mind, Body and World Together Again.* Cambridge, MA: MIT Press, 1998.

Clément, Raphaël. 'Stéphane Leduc and the Vital Exception in the Life Sciences', 2015, <https://arxiv.org/pdf/1512.03660.pdfarXiv:1512.03660>.

Clynes, Manfred E. and Kline, Nathan S. 'Cyborgs and Space', in *Astronautics*. Orangeburg, NY: Rockland State Hospital, 1960.

Cortázar, Julio. 'The Kings', tr. C. Svich. *Brooklyn Rail*, <https://intranslation.brooklynrail.org/spanish/the-kings-los-reyes/>.

Dawkins, Richard. *The Selfish Gene*. Oxford: Oxford University Press, 1976.

Drexler, K. Eric. *Engines of Creation: The Coming Era of Nanotechnology*. New York: Anchor Books, 1988.

Drexler, K. Eric, Smalley, Richard E,.'Nanotechnology: Drexler and Smalley make the case for and against "molecular assemblers"', *Chemical and Engineering News* 81:48 (2003), 37–42.

Driesch, Hans. *The History and Theory of Vitalism*. London: Macmillan, 1914.

Eisoldt, Lukas et al. 'Decoding the Secrets of Spider Silk'. *Materials Today* 14:11 (March 2011), 80–86.

Feynman, Richard P. 'There's Plenty of Room at the Bottom'. *Engineering and Science* 23:5 (1960), 22–36.

Fraenkel-Conrat, Heinz and Williams, Robley C. 'Reconstitution of Active Tobacco Mosaic Virus from its Inactive Protein and Nucleic Acid Components', *Proceedings of the National Academy of Sciences of the USA* 41:10 (15 October 1955), 690–698.

Freestone, Ian; Meeks, Nigel; Sax, Margaret and Higgit, Catherine. 'The Lycurgus Cup—A Roman Nanotechnology'. *Gold Bulletin* 40:4 (December 2007), 270–277.

Freud, Sigmund. 'New Introductory Lectures on Psychoanalysis. Lecture 33: Femininity', in *The Standard Edition of the Complete Psychological Works of Sigmund Freud*, ed. J. Strachey. London: The Hogarth Press and the Institute of Psycho-Analysis, vol. 22, 136–157.

Godfrey-Smith, Peter. *Other Minds: The Octopus and the Evolution of Intelligent Life*. London: Collins, 2016.

Gosline, John M. et al. 'The Mechanical Design of Spider Silks: From Fibroin Sequence to Mechanical Function'. *Journal of Experimental Biology* 202 (December 1999), 3295–3303.

Haraway, Donna. 'Promises of Monsters: A Regenerative Politics for Inappropriate/d Others', in *Cultural Studies,* eds. L. Grossberg et al. London: Routledge, 1992, 295–337.

———, 'A Cyborg Manifesto: Science, Technology, and Socialist-Feminism in the Late Twentieth Century', in *Simians, Cyborgs and Women: The Reinvention of Nature*. New York; Routledge, 1991.

Hardy, Bruce L. et al. 'Direct Evidence of Neanderthal Fibre Technology and Its Cognitive and Behavioral Implications'. *Scientific Reports* 10:1 (9 April 2020), 4889.

Hu, Xi et al. 'Biological Stimulus-Driven Assembly/Disassembly of Functional Nanoparticles for Targeted Delivery, Controlled Activation, and Bioelimination'. *Advanced Healthcare Materials* 7:20 (24 October 2018), 1800359.

Huang, Chuanhui et al. 'Effect of Structure: A New Insight into Nanoparticle Assemblies from Inanimate to Animate'. *Science Advances* 6:20 (13 May 2020), eaba1321.

Hurcombe, Linda M. *Archaeological Artifacts as Material Culture*. London: Routledge, 2007.

Ilday, Serim et al. 'Rich Complex Behaviour of Self-assembled Nanoparticles far from Equilibrium'. *Nature Communications* 8 (26 April 2017, 14942).

Irigaray, Luce. *Marine Lover of Friedrich Nietzsche*, tr. G.C. Gill. New York: Columbia University Press, 1991.

——, *Elemental Passions*, tr. J. Collie and J. Still. London: Routledge, 1999.

——, *The Forgetting of Air in Martin Heidegger*, tr. M.B. Mader. Austin, TX: University of Texas Press, 1999.

Japyassù, Hilton F. and Laland, Kevin N. 'Extended Spider Cognition'. *Animal Cognition* 20 (7 February 2017), 375–395.

Keller, Evelyn F. *The Century of the Gene*. Cambridge, MA: Harvard University Press, 2002.

Laplane, Lucie et al. 'Opinion: Why Science Needs Philosophy'. *PNAS* 116:10 (5 March 2019), 3948–3952.

Leduc, Stéphane. *La biologie synthétique*. Paris: A. Poinat, 1912.

——, *The Mechanism of Life*. London: Heinemann, 1911.

Lepore, Emiliano et al., 'Spider Silk Reinforced by Graphene or Carbon Nanotubes'. *2D Materials* 4:031013 (14 August 2017).

Levi, Primo. 'Quaestio de centauris', tr. J. McPhee. *The New Yorker*, 8/15 (June 2015).

Liu, Daniel. 'The Artificial Cell, the Semipermeable Membrane, and the Life that Never Was, 1864–1901'. *Historical Studies in the Natural Sciences* 49:5 (1 November 2019), 504–555.

Liu, Xiang Yang and Li, Jing-Liang. *Soft Fibrillar Materials: Fabrication and Applications*. Weinheim: Wiley-VCH, 2013.

Lo, Chiao-Yueh et al. (in press), 'Highly Stretchable Self-sensing Actuator Based on Conductive Photothermally-Responsive Hydrogel'. *Materials Today* 50 (2021), 35–43.

Maiti, Subhabrata et al. 'Dissipative Self-assembly of Vesicular Nanoreactors'. *Nature Chemistry* 8 (2 May 2016), 725–731.

Matsuda, Takahiro et al. 'Mechanoresponsive Self-growing Hydrogels Inspired by Muscle Training'. *Science* 363:6426 (1 February 2019), 504–508.

Matsumoto, Akira et al. 'Synthetic "Smart Gel" Provides Glucose-responsive Insulin Delivery in Diabetic Mice', *Science Advances* 3:11 (22 November 2017), eaaq0723.

Maturana, Humberto R. and Varela, Francisco J. *Autopoiesis and Cognition: The Realization of the Living.* Dordrecht and Boston, MA: D. Reidel, 1980.

Merleau-Ponty, Maurice. *Phenomenology of Perception*, tr. D.A. Landes. London and New York: Routledge.

Miller, Stanley L. 'A Production of Amino Acids Under Possible Primitive Earth Conditions'. *Science* 117 (15 May 1953), 528–529.

Mills, Douglas R. et al. 'An Extracellular Darwinian Experiment with a Self-Duplicating Nucleic Acid Molecule'. *Proceedings of the National Academy of Sciences* 58 (1967), 217–224.

Monod, Jacques. *Chance and Necessity: An Essay on the Natural Philosophy of Modern Biology*, tr. A. Wainhouse. New York: Knopf, 1971.

Morin, Edgar. *La sfida della complessità.* Florence: Le Lettere, 2017.

NASA. 'Life Detection', *Astrobiology at NASA.* <http://astrobiology.nasa.gov/research/life-detection/>.

Nature Nanotechnology. 'Nanotechnology Versus Coronavirus'. *Nature Nanotechnology* 15 (6 August 2020), 617.

Otake, Mihoko et al. 'Motion Design of a Starfish-shaped Gel Robot Made of Electro-Active Polymer Gel'. *Robotics and Autonomous Systems* 40:2–3 (31 August 2002), 185–191.

Patel, J. and Patel, A. 'Nano Drug Delivery Systems for Space Applications', in *Handbook of Space Pharmaceuticals,* eds. Y. Pathak et al. Basel: Springer Nature Switzerland, 2019.

Pennisi, Elizabeth. 'Untangling Spider Biology', *Science* 358: 6361 (20 October 2017), 288–291.

Plant, Sadie. 'The Future Looms: Weaving Women and Cybernetics'. *Body and Society* 1:3–4 (1 November 1995), 45–64.

———, *Zeros And Ones: Digital Women and the New Technoculture.* London: Fourth Estate, 1998.

Prigogine, Ilya and Stengers, Isabelle. *Order Out of Chaos: Man's New Dialogue with Nature.* Toronto: Bantam, 1984.

Ramberg, Peter J. 'The Death of Vitalism and the Birth of Organic Chemistry: Wöhler's Urea Synthesis and the Disciplinary Identity of Organic Chemistry', *AMBIX* 47:3 (November 2000), 170–195.

Rising, Anna. and Johansson, Jan. 'Toward Spinning Artificial Spider Silk'. *Nature Chemical Biology* 11:5 (14 April 2015), 309–315.

Schneider, Hans-Jörg (ed.). *Chemoresponsive Materials. Stimulation by Chemical and Biological Signals.* Cambridge: Royal Society of Chemistry, 2015.

Searle, John. 'Minds, Brains and Programs'. *Behavioral and Brain Sciences* 3:3 (September 1980), 417–457.

Shelley, Mary. *Frankenstein.* Ware: Wordsworth Classics, 1992.

Smalley, Richard E. 'Of Chemistry, Love and Nanobots'. *Scientific American* 285:3 (September 2001), 76–77.

Steven, Eden. et al. 'Carbon Nanotubes on a Spider Silk Scaffold'. *Nature Communications* 4:1 (10 September 2013), 2435.

Sufian, Mian M. et al. 'Safety Issues Associated with the Use of Nanoparticles in Human Body'. *Photodiagnosis and Photodynamic Therapy* 19 (September 2017), 67–72.

Tero, Atushi et al. 'Rules for Biologically Inspired Adaptive Network Design'. *Science* 327:5964 (January 2010), 439–442.

Traube, Moritz to Charles Darwin, Breslau, 2 March 1875 (Letter no. 9878), in *The Correspondence of Charles Darwin*, eds. F. Burckhardt et al. Cambridge: Cambridge University Press, 30 vols. (1985–), vol. 23.

Usmani, Sadaf et al., 'Functional Rewiring Across Spinal Injuries via Biomimetic Nanofiber Shelves'. *Proceedings of the National Academy of Sciences* 117:41 (September 2020), 25212–25218.

Von Bertalanffy, Ludwig. *General Systems Theory: Foundations, Development, Applications.* New York: George Braziller, 2015.

Wayland Barber, Elizabeth. *Women's Work, The First 20,000 Years: Women, Cloth, and Society in Early Times.* New York: Norton, 1995.

Wehner, Michael et al., 'An Integrated Design and Fabrication Strategy for Entirely Soft, Autonomous Robots'. *Nature* 536 (24 August 2016), 451–455.

Whitesides, George M. and Grzybowski, Bartosz. 'Self-Assembly at All Scales'. *Science* 295:5564 (2002), 2418–2421.

Wiener, Norbert. *Cybernetics: or Control and Communication in the Animal and the Machine.* Cambridge, MA: MIT Press, 2019.

———, *God and Golem, Inc.: A Comment on Certain Points Where Cybernetics Impinges on Religion.* Cambridge, MA: MIT Press, 1964.

Yarger, Jefferey et al. 'Uncovering the Structure-Function Relationship in Spider Silk'. *Nature Reviews Materials* 3:3 (March 2018), 18008.

Zou, Z. et al. 'Rehealable, Fully Recyclable, and Malleable Electronic Skin Enabled by Dynamic Covalent Thermoset Nanocomposite'. *Science Advances* 4:2 (9 February 2018), eaaq0508.

Index

2001: A Space Odyssey 91

A
abiogenesis 115–116
Agrippa, Cornelius 112
alchemy 85–86, 109–110, 112
Aldini, Giovanni 111
amino acids 25–26
Arachne 17–19, 19, 159, 171–172
artificial intelligence 51, 52–53, 56, 91
astronauts 141
atomism 105–106
atomic assembly 83–84
ATP 128–129
autocatalysis 125, 131
autopoiesis 78–79, 87, 98

B
Babbage, Charles 13, 160n
Barad, Karen 64, 163–165
Barber, Elizabeth Wayland 11
Bartlett, Stuart 125
Bateson, Gregory 105
Bénard cells 97–98, 99
Bénard, Henri 97
Bennett, Jane 165–166
Beverina, Luca 170n
biology vii, viii, 63, 115, 139
    synthetic 120–122, 127, 135, 140
    and chemistry 25–26, 77. 140
biomimicry 32
bodies
    and intelligence 57
    and nanomaterials 155–156
    cyborg 141
Bohr, Niels 164
Boltzmann, Ludwig 75

bottom-up
    vs top-down 67, 69, 86, 106, 69
brains 51, 53, 55, 56, 57, 58, 110
Brunetta, Leslie 33
buckyball 87–88
Byron, Lord 13

C
Cambou, Jacob Paul 30
carbon 153
carbon fibre 21
cells
    artificial 118, 119
cephalopods 51
Chalmers, David 57–58
chemistry 25–26, 88, 113, 114, 118
    and alchemy 112
    and materials science 5
    and biology 25–26, 77. 140
Chinese Room Argument (John Searle) 52–53
Clark, Andy 57–58, 62–63
Clynes, Manfred 141
cognition
    centralised view of 55
    embodied 58, 61, 62–63
    extended 64, 129
    in spider and web 149–151
complexity 75–77, 76, 79–80, 81
    and mind 103–104
complex systems 71, 78, 98–104
computation
    morphological 41
    Physarum 41–42
control 86–87, 89, 92, 106, 147–148, 171
cooperation 14
COVID-19 95, 170
Craig, Catherine 33
crystals 97, 118, 122, 136

culture viii, 12, 33, 64, 106, 126,
    151–152, 156, 160–163,
    165–166, 170
Curl, Robert 87
cybernetics 105, 135–136
cyborg 136, 138, 141–142, 162–163

D

Daedalus 169
Darwin, Charles 119–120
determinism 99
  and technology 18
disciplines, disciplinarity viii, 5, 140,
    148
dissipation, dissipative structures 98,
    99, 105, 125, 104–105, 129
DNA 25, 93
domination 64, 140, 142, 156, 161,
    163, 165, 170
Drexler, K. Eric 87–88
Driesch, Hans 69

E

ecology 19, 152, 169–170
Einstein, Albert viii
electricity 110–111
emergence 67, 76–77, 77–78, 78
energy 22, 23, 27, 32, 68, 92, 93,
    96–98, 110–111, 122, 125, 128,
    129, 149
  collateral 105
  consumption of 170
entanglement 164
entelechy 69
entropy 96
evolution 33, 119, 126, 132
  chemical 115
  of spider 149, 151
Ex Machina 91
experience 50, 57, 58, 61, 63, 76
extended mind *see* mind

F

far-from-equilibrium 104–105
femininity
  and technology 13, 160–161
feminism 162–163, 166
  and cyborgs 163
Fermi, Enrico 127
Feynman, Richard 82, 86–87
fluidity 161
form 41, 63, 103
  and life 117
Fox Keller, Evelyn 133
Fraenkel-Conrat, Heinz 93
Frankenstein, 110–113, 115, 138
Freud, Sigmund 159–160

G

Galvani, Luigi 110
genetics 95, 126, 132–133
genetic engineering 28, 95
glass 85–86
Godfrey-Smith, Peter 50
Golem 70–71, 79
graphene 33–34, 153
graphite 88
gray goo 91

H

Haraway, Donna 139–142, 143
    162–163
homeostasis 126
homunculus 109
Hurcombe, Linda 12
hybridity 6, 37, 138, 153–154, 170
  and the cyborg 163
  organic-inorganic 114
Hydra 37, 38
hydrogel 47–48
hysteresis 23–24

I

Icarus 171
identity 64, 103

indeterminacy 163–164
information 71, 105, 126, 131–134, 136
inorganic 109, 111
  and organic 72, 126
intelligence 14, 39, 49, 51, 58, 147, 148, 149
  and bodies 57
  and community 62–63
  and representation 147
  decentralised 31, 38, 43, 147
  non-human 50–51
  simulation of 56–57
intelligent materials 44, 49, 61, 62, 63, 148–149, 165
interaction 4, 67, 76
  and intra-action 64
interface 1–4, 4, 82, 141, 155–156
intra-action (Barad) 64
invariance 137
Irigaray, Luce 161–162
isomerism 113

J
Jacquard, Jean-Marie 13
Japyassù, Hilton 149

K
Kevlar 21
Kline, Nathan 141
Kroto, Harold 87

L
labyrinth 169
Laland, Kevin 149
Leduc, Stéphane 120–122, 127, 140
Levi, Primo 1
life 79, 100–101, 114, 115, 117, 127, 135
  alien 148
  and information 132–133
  and technnology 114
  definitions of 127–128

loom 13, 160
Lovelace, Ada 13, 160n
Lycurgus Cup 84–85
lyfe (Bartlett and Wong) 125–126, 135

M
machines
  self-replicating 71
materials
  bionic 154
  intelligent 49, 61
    and softness 62
  soft 34, 63
  smart 44, 51, 57
material culture viii, 12, 19, 34,
materials science 5, 13, 33, 122
matter 105
  and information 70–71
  chemical 114
  stickiness of 83
Maturana, Humberto 78
membranes 119
memory
  and intelligent behaviour 24
  and responsiveness 103
  in shape-memory alloys 45
  spatial, in spiders 150
  spatial, of *Physarum polycephalum* 41–42
Merleau-Ponty, Maurice 58–59
metabolism 98, 128
Miller, Stanley 116
mimetic polyalloy 138
mind 103, 105, 106
  and technology 156
  animal and material 151
  extended 57–58, 129, 149
Monod, Jacques 136–137, 142
monstrosity 140, 143
Morin, Edgar 76–77
morphogenesis 117, 137
morphology 93–94

mythology 17, 37
myths 69–70

N

nanocomposites 153
nanomaterials 46, 153–154
nanomedicine 95
nanoparticles 99, 155
nano-robotics 88
nanotechnology 33, 82–83, 84–85,
    86–89, 91, 91–92, 95–96, 100,
    106, 122, 170
nanotubes 153–155
natural selection 25, 33, 119–120
nature 32, 64, 140, 142–143, 153–156,
    160–161
  as 'artifact' (Haraway) 138–139,
    162–163
networks 4, 40–41, 151, 162, 164,
    166, 170
nylon 32

O

octopi 50–51, 61
Octobot 46
Oparin, Aleksandr Ivanovič 115
organic 114, 126, 140, 141, 162
organisms
  and machines 68–69, 72, 81,
    135–137
  and science 75–76
  artificial 109
  chemistry of 70
Orphan Drift 51
Osmosis 119n
Ovid 17

P

Paracelsus 109–110, 112
Pasteur, Louis 115
Perception 58, 63
phase shift 45
philosophy
  and science vii–ix

Physarum polycephalum 38–41, 43, 57
physics vii, 80
Plant, Sadie 31, 160
plasmon 85
polycephaly 37
polymeric materials 45–46
polymers 25, 32, 47, 96
  synthetic 63
polystyrene 99, 104
posthuman 62
Prigogine, Ilya 97, 98, 104–105
proteins 26, 45–46, 63, 93

Q

queerness 64

R

reduction 76–77
  and emergence 76–77
reductionism vii, 77
relationships, relationality 4, 14,
    80, 103, 140, 147–148, 163
    166–167
replication 83, 125, 128
  of viruses 93–94
representation 147
RNA 93, 95, 133–134
robots, robotics 49, 71
  soft 48, 62

S

scale 81–82
science 169
  and alienation 142
  and domination 140, 156
  and literature 1
  as unnatural 6
Searle, John 52, 55–57
segregation 128
self-assembly 30, 88, 92–96, 155
  static vs dynamic 92–93, 98
selfish gene 132
self-organisation 71, 78, 82, 98,
    103, 104, 106, 113, 122, 129,
    147–148

static vs dynamic 122–123
self-replication 131–132, 134
shape-memory alloy 44–45
Shelley, Mary 110–111
silk 20, 23, 25, 26–27, 30, 32–34, 83, 148, 151
    major ampullate 21–22, 26–27
simulation 56–57
skin 61
    artificial 49–50
slime 38–39, 41, 42, 43, 55, 64, 150
Smalley, Richard 87, 88–89
softness 34, 46, 51, 62
software 31
spiders 13, 17, 19, 20,21 23, 25, 28, 33, 149–150, 151, 153, 154, 159
Spiegelman, Sol 133
spontaneity 67–68
spontaneous generation 113, 114, 115, 120
Stengers, Isabelle 97–98, 104–105
structure 14, 113
subject 103–104, 164–165
supercontraction 27
surface 2–3, 82
surface plasmon resonance 84–85
surface tension 3
surfactants 3, 128–129
synthesis 5, 25–26, 110, 114, 140, 165
    vs representation 143–144
synthetic biology see biology
systems theory 77, 105

T
tactility 49
targeted drug delivery 94
technology 11, 17–18, 33–34, 64, 142–143, 171
    and domination 140, 161, 163
    and climate crisis 170
    and life 111
    and softness 12–13
    and the feminine 13
    and weaving 31

perishable 11–12
self-replicating 70
soft 31
teleonomy 137
Terminator 2: Doomsday 138
Thales 18
thermodynamics 68, 69, 96
    statistical 75
Theseus 171
three-body problem 75
time 123
    and life 117
    and organisms 68–69
tobacco mosaic virus (TMV) 93–94
top down
    vs bottom up see bottom-up
Traube, Moritz 118–120.

U
urea 113, 115
Urey, Harold 116

V
vaccines 95
Varela, Francisco 78
viruses 93, 96, 133–134, 170
vital force 111
vitalism 69, 120
von Bertalanffy, Ludwig 77

W
weaving 11–14, 17, 19, 26
    and women 160–161
webs 13, 19, 20, 21, 27, 150–151, 154
Wiener, Norbert 71–72, 135
Wiener (still) 72
Williams, Robley 93
Woese, Carl viii–ix
Wöhler, Friedrich 113
women 15
    and technology 11, 155–156, 159–161
    scientists 162
Wong, Michael 125